An Introductory Guide to Repairing Mechanical Clocks

Scott Jeffery MBHI

THE CROWOOD PRESS

Chapter 1

Introduction

In my first four years as a clock repairer I have served my apprenticeship under the watchful eye of Chris Baldwin CMBHI, achieved MBHI certification, run a profitable clock repair business and been commissioned to write this book. Getting to this stage has been challenging and inspiring, and my intentions with this book are to guide you through what I have come to see as the minimum of knowledge and understanding it takes to become a competent repairer of basic mechanical clocks in the quickest amount of time possible, while still allowing plenty of scope for future developments in the knowledge, experience and equipment it takes to become excellent in the field. I have done all I can to re-imagine my first year, and to write this book in a manner which I would have found easy to understand while being technically interesting.

My biggest challenge was to limit this book to the true beginner, to not delve too deeply into the many subjects of horology and, above all else, to keep it interesting.

Deep theory has always failed to hold my attention for long. The detail given to mathematical examples and long-winded explanations can make reading chapters on theory a real chore. To combat this I have kept the theory portion of this book on a need-to-know basis, but there is a lot to be said. I have opted to leave out any unnecessary mention of historical aspects in order to keep the words flowing and to keep the reader in a technical mindset. Photographs have been provided where I feel you could benefit from their presence, and the accompanying explanations have been written to work with them. In this way you are able to see the components being discussed as though you were sitting with me at the workbench.

My intentions are to make the reader aware of the existence of the many components and theories which they are likely to come across in their first year 'on the job' and not to provide a full in-depth study of each. With the understanding that things such as thermal expansion or circular error exist, you are able to make informed decisions. You do not need to be able to calculate it to perform a good-quality repair.

Being relatively new to the industry myself, I completely understand the position of the reader, and what some might see as a disadvantage I use to my advantage.

There are specialist tools for every job, but which ones do you actually need to get it done? Do you really need to invest in a full workshop, ultrasonic cleaning equipment and various lathes just to repair your great-grandmother's mantel clock? It is unlikely. I have put together a tier system to help you define your needs and buy the necessary level of equipment to suit. These are the 'three toolboxes'.

If you intend to turn this hobby into a business, full- or part-time, I highly recommend that you contact the British Horological Institute and enrol in their Technician-grade Diploma in clock and watch servicing, and get a qualification under your belt, wherever you are in the world. It can be studied as a distance learning course in your own time, as I did, and it will provide you with certification, industry contacts, a good reputation and all the help and support you could need. For those of you studying the course already, I hope this book will prove helpful in demonstrating the repair processes to you in a real-life scenario and in compiling the theory into a well-organized, readable and uninterrupted whole.

An early 20th century carriage clock, complete with original winding key and travel box.

The front plate of a modern Westminster chiming longcase.

For more information on some of these repair processes, video tutorials, gear train calculators, or just to say hi, I can be reached by visiting www.learnclockrepair.com or www.hampshireclockworks.co.uk.

Enjoy!

Scott Jeffery, BSc, MBHI

Chapter 2

The Clock Repairer's 'Three Toolboxes'

I have split the relevant tooling of my own workshop into three separate categories, or 'toolboxes'. These three toolboxes are intended for:

- the beginner who wants only to make small repairs, and perform routine maintenance
- the enthusiast who intends to complete the majority of their own repairs and those of friends
- the aspiring professional, who intends to undertake repairs for profit.

For every process demonstrated and explained within this guide, I note which toolbox you will need in order to complete it to a high standard. If the job requires tooling which you have not yet acquired, I suggest you buy just the tools needed and slowly build up to the next toolbox. However, if it is a once-in-a-lifetime repair for your interest level, outsource it to a competent, qualified professional who can be found at www.bhi.co.uk.

TOOLBOX 1: THE ABSOLUTE MINIMUM

This toolbox will help you to keep your clock running and extend the period between services, and ultimately keep the costs of those services down. For minimum expenditure, these tools will allow you to regularly oil your clock, repair common breakages and make adjustments to correct the more common issues which may arise. Eventually the time will come when pivots are worn and you will have the choice of either taking your clock to a repairer for overhaul, or upgrading to Toolbox 2 and doing it yourself.

The tools I recommend here are the ones I use most often. I have not suggested makes or names except where offering details of my own equipment because this is not an important part of the choice. What is important to remember is that many of the clocks you will encounter were made two hundred or more years ago, with basic handmade tools which were properly sharpened and well maintained. Provided you use them correctly and with care, the cheapest tool can last a lifetime. Abuse them, and even the named brands will not hold up.

Some tools are luxuries and can simplify or speed up a specific job; these are reserved for Toolbox 3, an advanced collection for the aspiring professional.

There are many tools available which are unnecessary or easily made at home; pivot locators and beat-setting tools, for example. I do not recommend buying these.

The majority of this tooling can be bought at auction for a fraction of the retail price, and can easily be refurbished. This is how I have gathered most of my tooling, and if you are confident in reconditioning your own tools, it is the method I recommend for any beginner.

Screwdrivers

For clock work you will need a good selection of screwdrivers. Platform escapements use screwdrivers as small as 0.3mm, while for longcase clocks you can often use screwdrivers more often associated with DIY. In fact, some of my favourite screwdrivers were bought from a DIY supplier, although I had to thin the blades to suit my needs.

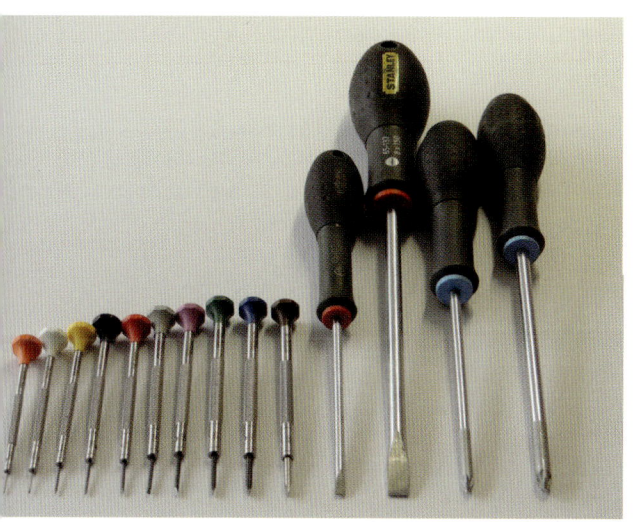

Suitable screwdriver set for clock repair.

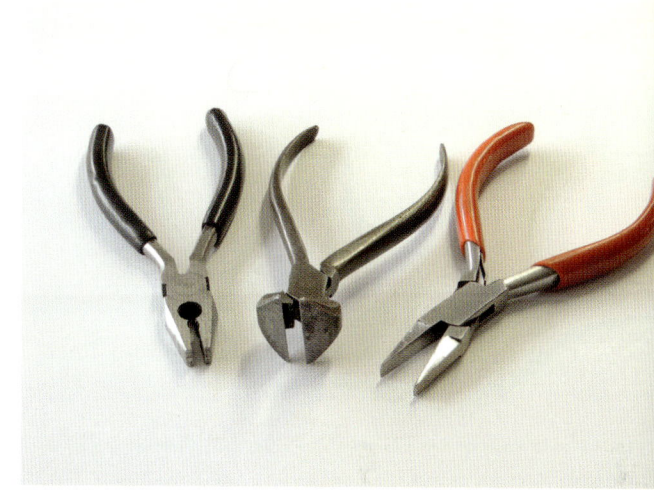

Combination, end cutting and smooth-jawed pliers.

I recommend buying a ten-piece set of clock- or watchmaker's screwdrivers from a horological supplier, making sure it includes the larger 3mm blade which is most commonly used in clock work. Choose a set with interchangeable blades as they will chip and require sharpening or replacing, and a swivel top is a nice feature for comfort. For larger work, I use a 'Stanley Fatmax' twelve-piece set. This gives a good selection of standard and Phillips screwdrivers. Phillips screwdrivers do make appearances in modern clocks so having a few sizes to hand is useful, although they should not be used in antique work. The blades of the standard 'flat' screwdrivers are generally too fat for clock work, but they file down nicely and last a long time.

Refurbishing your screwdriver blades can be done with a sharpening stone or a file. A honing guide can be used to maintain the angle. To sharpen by hand, hold the flat side of the blade down on the stone or, if using a file, support the screwdriver on the edge of the bench and apply the file to the blade. While maintaining the angle, draw it back and forth until a flat surface is produced. Turn the blade over and repeat the process on the other flat until the two sides are even. Finish by holding the blade upright on a sharpening stone and stroking a few times in a circle; this final step removes the weak tip which is liable to chipping.

Pliers

A good selection of pliers is essential in clock work, but for a few basic jobs you can get away with owning just the following:

- Combination pliers: the grooved jaws are good for pulling pins, and the cutting part for shortening them. Combination pliers will keep the bill down but are not always the best option.
- End cutters: a set of end cutters are my most versatile pliers in the shop. They can be used to cut pins, they can be rocked against the plate (protected by a piece of scrap material) to pull out tight pins and, because the jaws remain parallel as they open and close, they can be used safely to loosen nuts. I have a blunted and polished pair for just this purpose.
- Smooth-jawed pliers: a pair of smooth-jawed pliers is necessary for manipulating and shaping soft components; the smooth jaws protect the metal from damage caused by a serrated jaw.

To maintain your pliers, regularly remove any burrs and sharp edges on them with a fine file. If they do not take to filing, as hardened steel will not, use a stone or diamond lap to produce the same result.

Soft Solder, Soldering Iron and Flux

With soft solder you can repair a number of components, fix loose or sloppy components and refinish pallet faces. Although the excessive use of soft solder is frowned upon, it is useful for small repairs. I use soft solder in certain repairs and it is perfectly acceptable when done well.

You will require a supply of lead-free solder and a good flux to help it flow to where you want it. You will also need a high-power soldering iron or an attachment for your microtorch, as you are generally heating up larger components than standard electrical soldering irons are designed for. Smaller soldering irons are useful for a few repairs but are rarely used. If your budget stretches to it, a small blowtorch is recommended for when a soldering iron is not up to the job; it will also allow you to harden, temper and anneal your materials.

The tip of your soldering iron should be kept clean in order for it to work effectively. To achieve this, wipe the hot end on a damp sponge to clean off flux and excess solder. The tip is best preserved when 'tinned' with solder after every use.

Ball-Pein Hammer

A ball-pein hammer of about 110 grams (4oz) to 170 grams (6 oz) is just right, although hammers are generally measured and sold in oz and the size is down to personal preference. I personally use a small 4oz hammer with a short handle of about 15cm (6 in). This gives me a good 'whack' when needed but gentle controlled taps most of the time. The ball-pein hammer is good for riveting and stretching metals, as well as for providing blows to accurately placed stakes and punches. Any dents and marks on the hammer head will be transferred to the work piece when you hit it, so periodically polish the working surface for best results.

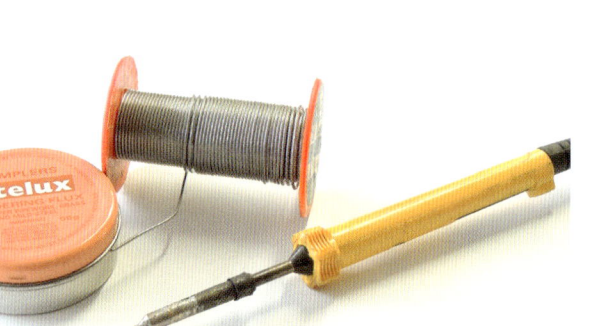

Soft-soldering equipment.

Ball-pein hammer for general work and riveting.

Needle Files

A good selection of needle files is not absolutely necessary at this stage, but it is recommended. If you are soldering, then they are useful for cleaning the materials back to bare metal beforehand, and for removing excess solder afterwards. I use an old square needle file clamped in a vice for rotating the collets in modern minute hands, for making sure that they strike dead on the hour. They can be used for removing burrs and wear grooves in some components, and can be used for reconditioning your other tools as they wear and chip.

To maintain your needle files, periodically remove any material stuck between the teeth by running the tip of a razor blade through the grooves. Applying chalk to the teeth before use helps prevent this build-up, whilst filing soft metals like solder increases the 'clogging' of the file teeth.

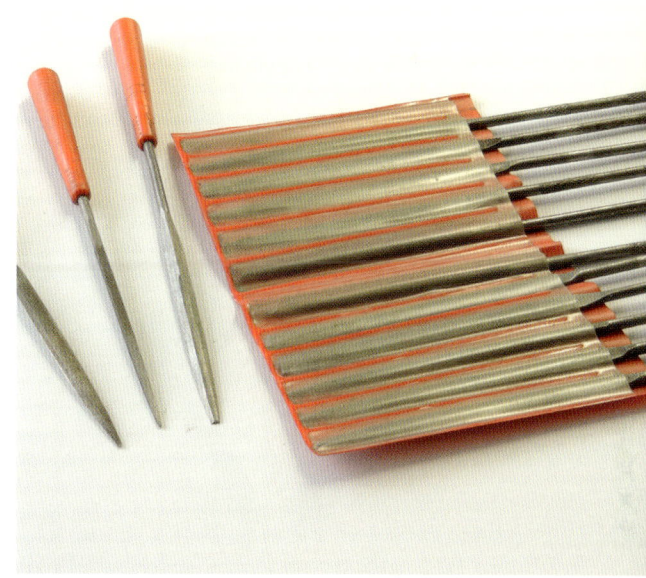

Selection of needle files.

Clock Oil and Oiler

You should be using a good-quality clock oil for maintaining your clocks, such as those sold by horological suppliers. When oiling platform escapements you should use a medium watch oil. For a clock oiler you can use a piece of brass wire, with the end hammered flat and filed into a diamond shape. For oiling watches I would suggest buying a commercial oiler. I prefer a pen-type oiler which dispenses oil with the click of a button.

The oil pot and bottles should be kept out of direct sunlight and dusty areas to prevent deterioration of the oil, and the working supply should be kept clean and periodically refreshed.

Movement Holders and Test Stands

You are going to need to test your clocks thoroughly after repairing them, and various test stands will be needed for the many types of movement you will be working on. Plans for building various useful test stands can be found in the resources chapter; alternatively you can buy commercial test stands if you decide not to make them yourself.

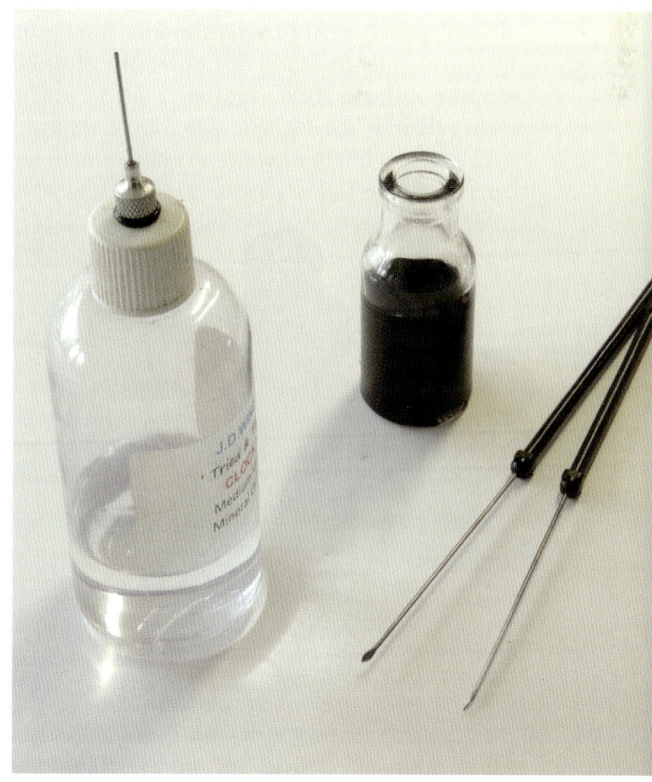

Clock oil and oilers.

Mainspring winder.

I regularly use when overhauling movements, from basic timepiece carriage clocks all the way up to musical Victorian longcase clocks. This toolbox does not include specialist tools like the truing callipers or the watchmaker's lathe; these will be part of Toolbox 3.

You will still be limited on component manufacture at this point, and some of the more complicated repairs will be beyond the capabilities of your equipment and experience. However, 80 per cent of jobs which pass through our workshop could be completed using these tools alone; where we choose to use more advanced tooling is usually for speed and cost-effectiveness.

Mainspring Winder

When you start disassembling clock movements it is absolutely necessary to remove and inspect the mainsprings. This is best done with a mainspring winder. Using the mainspring winder helps to avoid distortion of the mainspring, and is by far the safest way of removing and fitting strong springs such as those used in fusee clocks. There are multiple designs of mainspring winder and you should pick the one you find most comfortable to use. My choice of mainspring winder is, in my opinion, the simplest type and the one I highly recommend. Instructions for its use and photographs can be found in the later chapter for reassembling the movement.

Pivot File and Burnisher

This is used for correcting and work-hardening worn pivots. As oil dries and gathers dust, the pivots begin to wear and become pitted and grooved. If this is not corrected when the movement is cleaned, there will be excessive friction in the pivots and accelerated wear to the pivot hole. The pivot file and burnisher is often a double-ended tool. The file end is an extremely fine file and cannot be substituted

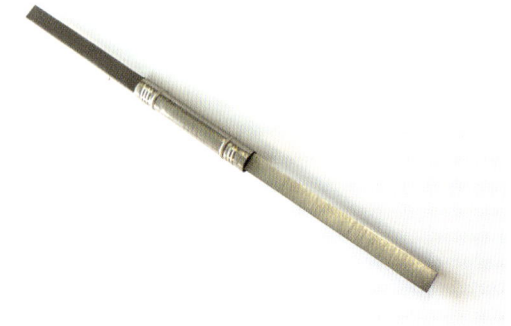

Pivot file and burnisher.

for a needle file, while the burnisher is a flat, hardened piece of steel, lightly grained across its width. The burnisher has a sharp edge and a rounded edge to suit different pivot types. Be sure to use the correct edge in contact with the shoulder in order to ensure no material is left in the corner which would cause the arbor to bind in its pivot hole.

Maintaining the burnisher is done by refreshing the cross-grain periodically on a medium stone or emery paper stretched over a piece of wood.

Selection of hand chucks and vices.

Pin Chucks/Hand Vices

A selection of pin chucks and hand vices is necessary for holding small components and tools; they are also used when correcting worn pivots if a lathe is not available. Buy a graduated set which will serve most purposes.

Cutting Broaches

Cutting broaches are used for opening pivot holes to an exact size, either to fit a bush or to accept a freshly burnished pivot. The broach itself is a long tapered piece of hardened steel, ground with five sides. The points at which the sides meet form a cutting edge. The broach is inserted into the hole and twisted. It is best to attach a handle to your broaches, or use a pin chuck. Maintenance of the cutting edge is by stroking the five flats on a stone or diamond lap, ensuring that their angle is not changed. Good-quality broaches have a reliable taper and will last a lifetime if treated well.

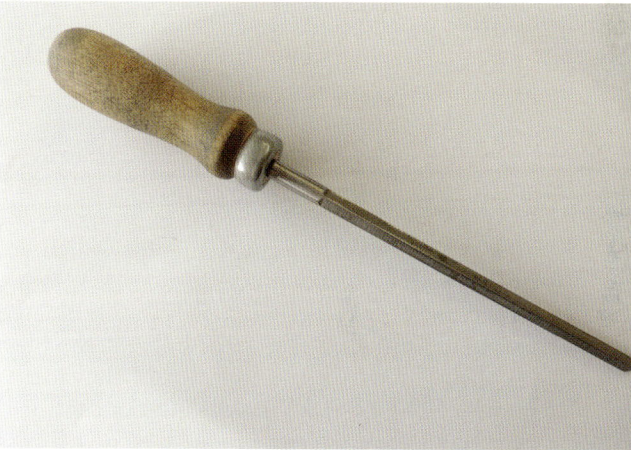

Large cutting broach.

Smoothing Broaches

Smoothing broaches are used after the cutting broach and are hard steel tapered rods which are ground perfectly round. As a smoothing broach is twisted in a pivot hole with a small drop of oil, it burnishes or work-hardens the surface of the brass.

A broach can also be used for checking a pivot hole for wear by inserting it into the hole, which should be as round as the broach is. A worn pivot hole will show a slither of light to one side of the broach.

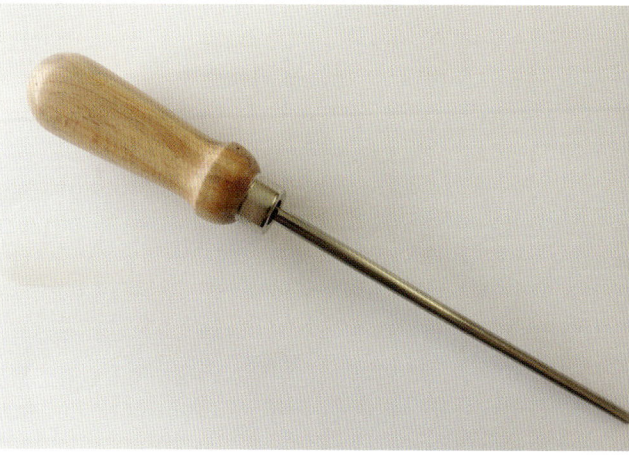

Large smoothing broach.

Good selection of brass bushes for clock repair.

Brass Bushes or Bushing Wire

Brass bushes are consumable items which are used to correct badly worn pivot holes. A wide selection should be purchased and graduated sets are available. The outer edge of the bush is slightly tapered to match that of the cutting broach. The inner hole is perfectly concentric to the outer edge and should be opened and burnished with your broaches. Brass bushes can be made in the lathe although much work is saved if a good-quality set is purchased.

Oil Sink Cutting Tools

Oil sinks are needed to retain the small drop of oil placed on the pivot holes. When you bush a worn hole the original oil sink will be cut away and you need to put it back. Oil sink cutters are readily available from suppliers and look like miniature pizza cutters. The round cutting blade is placed on top of the pivot hole where it will centre itself, and rotated in the fingertips. A round dimple will soon appear which will eventually become the oil sink. I use a modified drill bit in which the end has been shaped and polished to produce a good-quality oil sink.

Stakes and Anvils

Riveting punches and a staking block or anvil are needed for riveting large bushes. In fact, they serve so many purposes that it is highly recommended to just buy a good-quality staking set as listed in Toolbox 3. For this toolbox, simple domed and flat punches and a block or small anvil will suffice.

Selection of stakes and anvils.

Machine vice and soft-jawed bench vice.

Machine Vice and Bench Vice

A machine vice makes a very handy bench companion and can be used for a huge variety of applications, from holding components for filing or being used as an anvil, to acting as a temporary movement holder. A small bench vice is also handy; I use mine for holding a block of hard wood for refinishing pivots, closing tight spring barrels and holding mainsprings when remaking the hooking eye, etc.

Bristle Brush and Chalk

Chalk-brushing is the traditional method of adding a final finish to clean clock components; the stiff bristled brush is drawn over the chalk and then used to scrub the plates thoroughly. Often the plates will take a polish from this

Stiff-bristled brush and chalk for cleaning components.

action and all traces of cleaning residue will be removed. The chalk will absorb any remaining cleaning product as well as act as a light abrasive to polish the brass work.

Scratch Brush

A fibreglass scratch brush is a small pen-like tool which has a fibreglass tip. The fibreglass provides a sharp scrubbing action which moulds itself to the situation in hand. Scratch brushes are indispensable when it comes to cleaning pinion leaves and other tough-to-reach areas. As the tiny glass fibres break away they are liable to become stuck in your fingers, so I recommend using latex gloves or finger cots and to sweep the bench afterwards. The fibreglass tips do wear but refills are available.

Fibreglass scratch brush.

Pegwood

Pegwood can be purchased from horological suppliers in various diameters. These slim wooden rods have a tight grain which adapts well to being sharpened to a fine point. This point is inserted into a pivot hole and twisted, re-sharpened, and the process repeated until it emerges from the hole spotlessly clean. 'Pegging out' pivot holes is an important part of the cleaning process, and pivots which are not pegged out are certain to wear sooner than those which are. At a pinch, a cocktail stick will work well.

Small selection of pegwood.

Bench Knife

A good knife will be needed for sharpening your pegwood, but you will find many other uses for it. A sharp pocket knife will do.

Diamond Files

Diamond files are synthetic diamond-impregnated files which work excellently in situations where a steel needle file will fail to cut. I use diamond files for remaking and repairing the hooking eyes of mainsprings, a job which will destroy steel files in no time.

Two types of blade for use at the workbench.

The Clock Repairer's 'Three Toolboxes' 19

Diamond needle files, excellent for filing hardened steel.

Tin snips, for cutting sheet metal.

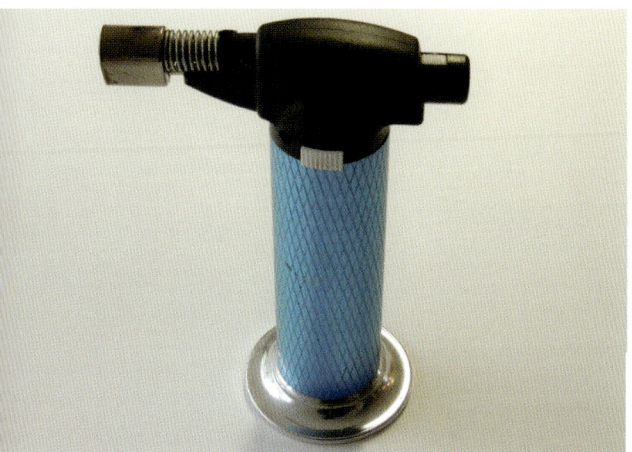

A micro blowtorch, for everything from soldering to heat treating.

Tin Snips

These are used for trimming away the torn end of a mainspring in preparation for making a new hooking eye, although you can just as easily snap the excess spring away as the steel is hardened and brittle, or just replace the mainspring entirely. These are not a necessary purchase, but are handy.

Blowtorch

A blowtorch will be used for many jobs: annealing materials before straightening bent pieces, hardening and tempering, soldering, bluing, etc.

TOOLBOX 3: THE ASPIRING PROFESSIONAL

The aspiring professional's toolbox is essentially a cut-down list of the tooling in my workshop. In addition to Toolboxes 1 and 2, Toolbox 3 will include lathes, depthing tools and cleaning machinery, etc. You will need a large workshop to house this level of equipment. However, with these tools, a little experience, and the knowledge shared within this book, you will be ready to go professional and produce good-quality work economically.

Watchmaker's Lathe

Your lathe will quickly become the most valued tool in the workshop. With it you will be able to re-pivot arbors, turn balance staffs, file and burnish pivots quickly, bush barrels, and make screws and true eccentric wheels, etc. There are so many lathes out there that it can be very difficult deciding what you need.

Any complete 8mm watchmaker's lathe will satisfy most of your needs: they are excellent for turning by hand and can be picked up second-hand for a reasonable price. Accessories are widely available at auctions and online, while some manufacturers still make new collets and tailstock accessories. I have found some limitations due to the small size of this lathe so I have a second one in the workshop.

An 8mm watchmaker's lathe.

The 10mm Pultra serves all the likely needs of the clockmaker. When well equipped you are able to do all the work of the 8mm lathe, but you have the ability to work to a larger scale. I use my 8mm BTM lathe for hand-turning and making small components as well as for re-pivoting arbors because I like its small and unimposing stature for up-close work. For all other jobs I prefer the 10mm Pultra lathe, which works nicely with the sturdy cross slide and drilling tailstock for boring holes and truing wheels.

With the right accessories any lathe can be made to suit your needs, but for the aspiring clockmaker I recommend starting with an 8mm lathe, and when this begins to show its limitations upgrade to a 10mm lathe as I did. Buy the most complete lathe possible; a full set of collets and step chucks is highly advisable, and three- and four-jawed chucks are a huge advantage for some work. A drilling tailstock is advantageous but you can do good work without it provided you have a tailstock centre to guide you when drilling by hand.

Depthing Tool

Every time a worn pivot hole is bushed, it is important to ensure that the pivot is put back to its original centre point. Unfortunately this is often overlooked and, with time, the pivot hole wanders away from its original centre. Eventually the depthing between a gear and pinion pair becomes so bad that the clock will stop for

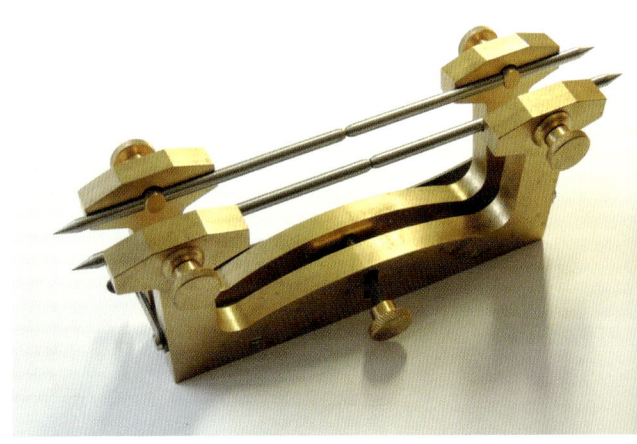

Clockmaker's depthing tool (large).

seemingly no reason. When this happens you need to check the depthing of all the wheels and pinions to see where the problem lies and to do this you need a depthing tool.

Depthing tools allow you to set up the wheel and pinion so that you can adjust their depths with micrometer accuracy. By experimentation and a good understanding of the theory you can find the correct position for the pivots and mark the clock plates with the scribe for drilling.

Truing Callipers

A pair of truing callipers can be used for both truing and poising any small wheel, such as the balance and escape wheels encountered when working on platform escapements. This will not be your most frequently used tool but they cost very little and can be extremely useful in diagnosing problems with platform escapements.

Jewelling Tool

The jewelling press has many uses besides removing and fitting press-fit jewel-holes. It is similar in design to a staking set, but at the top it has a micrometer depth stop and a lever to gently press the tool into action. The jewelling tool comes with a number of reamers used to open a hole and produce a perfect fit for its jewel. The reamer is then replaced with a pressing tool to press the jewel into the hole with no danger of damage and to ensure absolute parallelism. The depth to which the jewel is pressed into the plate is controlled by the micrometer gauge, thus giving excellent control of end-shake to the wheel.

The jewelling tool often comes with accessories for tightening cannon pinions, reducing holes in small hands and for use as a micrometer, and I often use mine for accurately bushing small pivot holes in platform escapements.

Rubbed Jewel Tool Set

Rubbed-in platform jewels can be replaced by manipulating the setting using these tools. They are cheap enough second-hand if you can find a set, which makes it possible to repair antique jewel-holes. There are three sizes of the open-

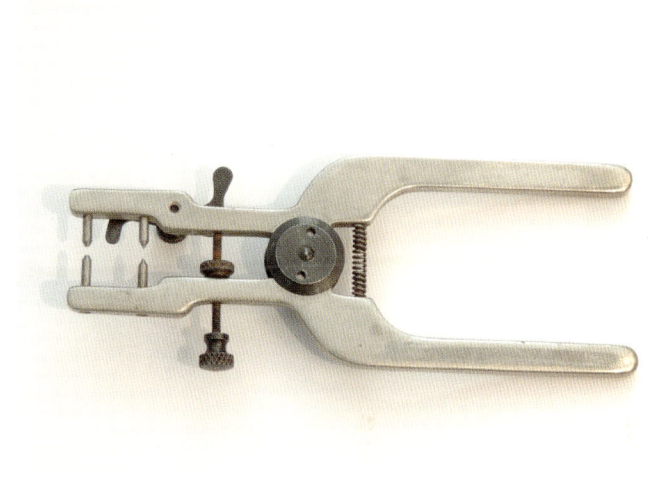

Pair of truing callipers.

Jewelling press.

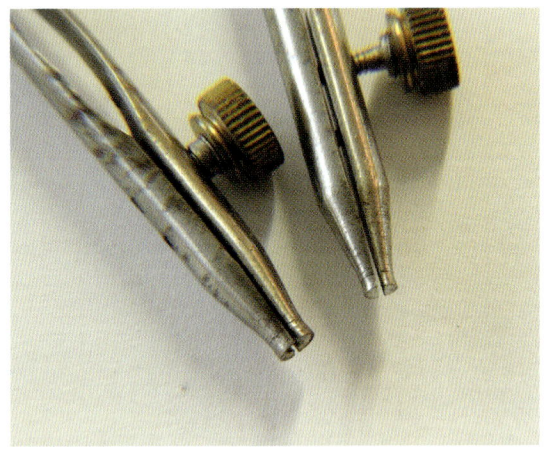

Set of jewelling rubbers.

Comparison of closing and opening jewelling rubbers.

ing and three sizes of the closing tool. Their use is covered in the repairs chapter.

Staking Set

A good staking set has a huge variety of uses and should be something you aim to buy relatively early on in your career. The staking set is an improvement over the riveting stakes and anvil from Toolbox 1 because it provides a method of holding a punch vertically and acts as a third hand, so you can use one hand for holding the component and the other for holding the hammer. It also provides either a hole or specially designed stake on which to position the component to be riveted, pressed, stretched, closed, driven or punched.

Staking sets are used when fitting a balance staff, driving out broken screws, reducing or reshaping holes, riveting small wheels to pinions, removing and replacing rollers and balance springs, and so much more.

Complete staking set.

To use the staking set, first take the conical tool and place it in the frame; this will centre the hole in the staking block with the punch guide-hole. Lock the staking block with the screw or lever on the rear of the tool and replace the punch and stake with the correct tool for the job. The punches can be either pressed by hand, or tapped with a small hammer.

Ultrasonic Cleaning Tank

Although not strictly a necessity, an ultrasonic cleaning tank will greatly increase your efficiency and turnaround time, as well as drastically reduce the effort needed to clean intricate components properly. I have added this to Toolbox 3 because it can be a rather expensive investment intended for the aspiring professional. When cleaning with an ultrasonic tank, a process which I describe in full later on, you submerge the components fully in cleaning fluid and the ultrasonic tank creates thousands of tiny implosions which blast the soft grime away from the solid metalwork of the movement which is simultaneously degreased by the cleaning agents in the fluid.

Ultrasonic and watch-cleaning machines.

Watch-Cleaning Machine

Rotating watch-cleaning machines are excellent for cleaning small components such as platform escapements because the cage system and gentle agitation of the cleaning fluid ensures that no damage comes to the parts while they are cleaned. Often they include a drying chamber into which the cage is placed for several minutes after the fluid has been rinsed off.

Micrometer and Callipers

Accurate measuring is essential when working with such tiny components, especially when measuring to order new springs. Clock mainsprings are often as thin as 0.25mm and are graduated in stages of 0.05mm so trying to use a ruler or 'eyeballing it' does not work. Vernier callipers and micrometers are best bought in millimetres rather than inches in the UK as this is how new components are labelled and supplied.

Measuring equipment for everyday use.

Pillar Drill

A small pillar drill is essential when you start manufacturing components. There are many manufacturers of micro-drills, but provided that there is some form of speed control and some 'feel' to the feed arm, any model will do. Accuracy is extremely important so it is best to measure the eccentricity of the chuck provided by fitting a steel rod which is known to be perfectly round and straight (a ground carbide drill shaft is good for this), and setting up a dial gauge to rest on this rod as near to the jaws of the chuck as is possible. Zero the dial and rotate the drill chuck by hand. The dial gauge should show no significant deviation as the rod rotates. If your chuck is showing eccentricities, rotate it until the lowest reading is obtained; at this point the rod is furthest from the gauge. Mark the chuck with a marker pen in line with this point and remove it from the drill following the manufacturer's instructions. The mark you made is the highest point on the three jaws of the chuck, which is pushing the rod out of centre. Use a small diamond file and reduce the jaw (or jaws if the line falls between two) nearest the mark and retest. Continue to repeat this process until concentricity is achieved.

Small pillar drill.

Files and Stones

You already have a selection of needle files and diamond files from Toolboxes 1 and 2, but when making components (as will be possible with Toolbox 3) you will need some larger files suited to more general engineering work. Buy a selection of new files in the 10cm (4in) and 15cm (6in) sizes; second cut is good for clock work. Half-round files are very useful in the smaller 10cm size for crossing out large wheels, and large flat files are useful for flattening the surfaces of components. You will need a variety so I recommend buying a set. Fit handles to all files over 10cm (4in) as the provided tang is not usually long enough to use as a handle (as with needle files).

Various stones can be very useful; my favourite ones are diamond laps of medium and fine grade. Modern synthetic stones are generally better than the old-style oil and Arkan-

Good selection of files and stones.

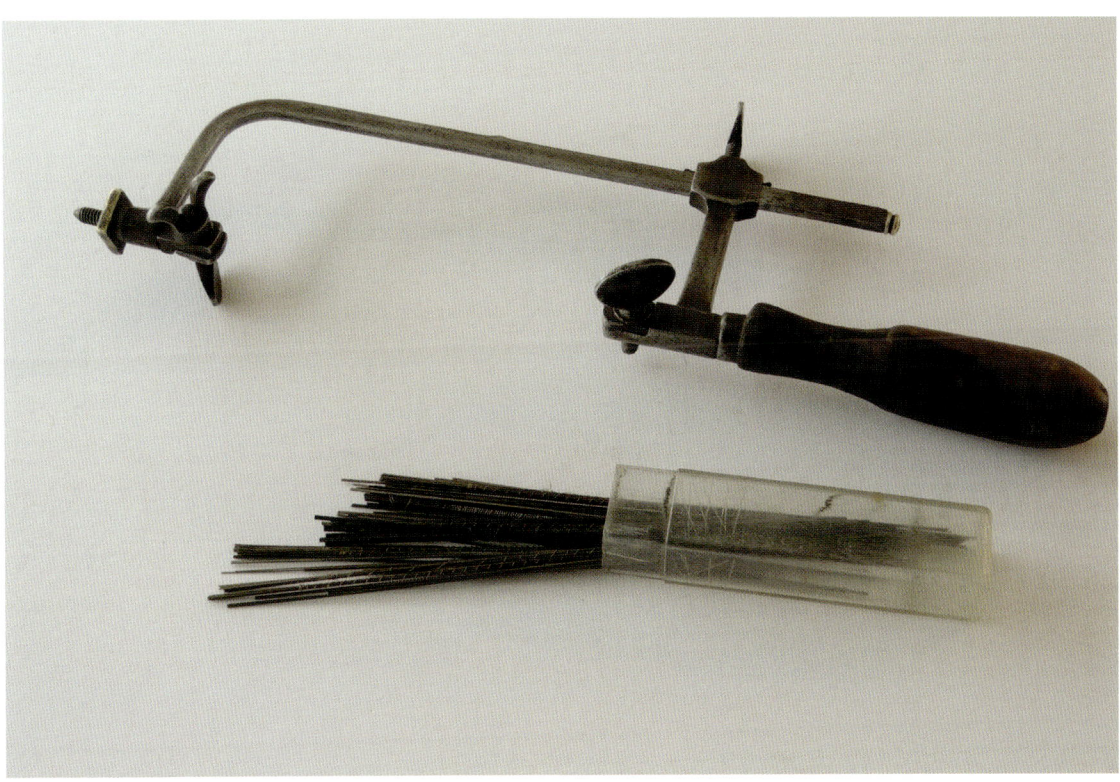

Piercing saw with a selection of blades.

sas stones which chip easily and pit with use. Diamond stones or 'laps' come in a variety of grades and make excellent general-purpose grinding and sharpening stones; they are used dry and cleaned very occasionally with water. As the 'diamond' coating is bonded to a metallic base, they do not become pitted with wear or chip when dropped and they last for a very long time, becoming slightly finer as the coating wears down. If you find that shaped stones would be a useful addition to your toolbox, I would recommend Degussit stones which are extremely hard and hold their shape; the finer grades provide a good finish, whilst the coarser grades can be used almost like a file.

Piercing Saw

When making components you will need to use a piercing saw to rough out intricate shapes, and follow that up by filing to size and stoning to a good finish. Piercing saw frames come in a variety of depths, but for the clockmaker piercing mainly small components, a 7.5cm (3in) frame should be adequate. The blades are inserted into the frame so that they cut on the pull stroke rather than the push stroke. This stops the frame from flexing which would result in the blade going slack and breaking.

The saw blades themselves come in a variety of grades, ranging from 8/0, 7/0 and 6/0 to 1/0, 0, 1, and so on up to 8, with 8/0 being the finest, with the highest number of teeth per inch, and 8 having the largest teeth. Selection of the correct saw blade for the job is not hugely scientific; for thin materials, use a finer blade, aiming as close as possible for a minimum of three or four teeth 'within' the thickness of the material at any time. A smear of beeswax on the blade or a drop of clock oil helps the blade cut freely.

Hacksaw

A hacksaw and junior hacksaw are useful for quickly removing the bulk of material to be pierced or for cutting a rough blank for making

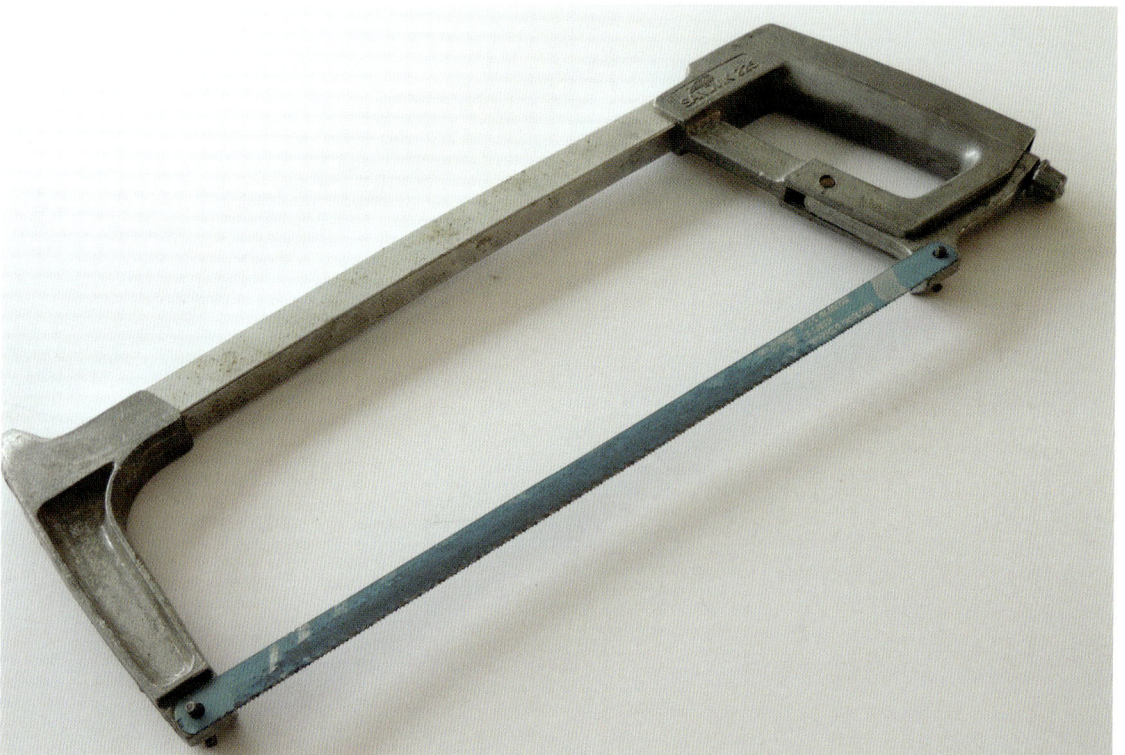

Large hacksaw.

a component. Hacksaw blades are fitted to cut on the push stroke and are measured in teeth per inch (tpi): 32tpi is good for most situations in clock work.

Taps and Dies

A good selection of taps and dies can be expensive but is worth every penny. BA sizes are useful in clock work so I would recommend starting with a set of BA taps and dies, and add to it over time with the metric taps and dies which are now used by major manufacturers. I find my most commonly used taps and dies to be 5BA and 9BA. They come in carbon steel (CS) or high-speed steel (HSS).

To maintain the condition of your taps and dies, advance the cutter forwards one turn, then half a turn back to clear the swarf and break the chips of material as it cuts. You should use a good cutting oil which will both preserve the cutting edge and prevent rust during storage.

Examples of taps and dies with their holders.

Oxy-propane micro-torch for an intense heat source.

Silver Solder

This is the missing link between soft solder (electrician's solder) and brazing. The required heat source will be more powerful than that used for soft soldering, and the flux will also be different. Silver solder is used in situations where structural integrity is necessary and does require some practice to get good results. It comes in grades rated by the temperature at which it melts. This allows you to solder two or more components in close proximity without melting the previous joint. For example, start with a high temperature solder for joint 1, and then use a low temperature solder for joint 2. Joint 1 will then stay solid while the soldering of joint 2 takes place.

Smiths Little Torch

This is an excellent oxygen/propane micro blowtorch which produces brilliant results when silver soldering due to its intense high-quality flame. My abilities in soldering grew instantly upon purchasing this bit of kit.

Chapter 3

Workshop Design

Your workshop requirements will depend on how involved you intend to become. For those of you who want to maintain your own clocks between overhauls and keep your bills down, you need nothing more than a clean desk, a kitchen sink and somewhere to keep your tools. At the other end of the spectrum is the aspiring professional for whom I recommend a custom-designed workshop, set up for maximum efficiency of the space. A spare bedroom or large garden shed will work, but you could consider a rented workshop to provide a professional front for dealing with customers.

When designing your workshop you should consider any inconveniences caused to those around you, especially regarding parking and deliveries at the inconvenience of your neighbours. If you cannot provide adequate parking, you should consider collecting and delivering where possible. If you are working with the public you will require a higher level of insurance to cover any potential injury or damage caused on site whether at your premises or theirs, this is called public liability insurance and will cover you for most conceivable interactions with the public. Your contents insurance provider should be able to provide public liability insurance, a package deal is the cheaper option. Your car insurance should also be upgraded to cover the transportation of customers property.

As your skill and understanding increases, so will your tooling and the workspace required to store and use it. You will soon have a lathe (or two), a pillar drill, a large bench vice and several clocks on test at any time. For maximum efficiency you will need the most commonly used tools to hand, including your lathe and a micro-drill of some form. Your workbench should be a comfortable height for you to work at when seated. You will benefit from making a small bench-top table to raise the worktop up to a height just below your collar-bone: this makes putting the plates of a movement back together much easier by raising them to eye level. I also recommend making a raised table for your lathe, so that you can look down on the work piece at roughly a 45-degree angle without having to hunch. Eventually you will be turning balance staffs and re-pivoting at this lathe, during which you will spending up to a couple of hours at a time in the same position. Plans for these tables can be found in the resources chapter at the end of this book.

You will need a large workbench. When you are working on a longcase clock you will need space to set aside the 30cm (12in) dial and seat board, a place to store the components as you strip them, space for your tools and at least a 30 × 30cm (12in) square of clear, uncluttered space for performing repairs. Bear this in mind when buying or building your workbench. I recommend that your workbench has a sacrificial surface which can easily be replaced. My bench has a 120 × 60cm (4 × 2ft) top for the main working area: this is a standard size for sheet hardboard and plywood so I can simply lift it off and fit a new piece periodically.

Natural light is by far the best if available; otherwise, well-placed electric light will do a reasonable job. Daylight bulbs produce a simulated natural light which is kinder to the eyes and easier to work under. Place a well-diffused light high up above the workbench to avoid strong shadows and have a small bench lamp for when stronger light is needed.

General safety should also be considered in the design of your workshop. The majority of your electric tooling will most likely be bought

My workbench.

from auction, and date from decades ago, it is wise to re-wire where possible, adding earth wires to the metal frames of the machines. Electrical equipment should be located away from water sources. It would be wise, if you are building your workshop from scratch, to add an isolation switch to cut off the power supply to the entire workshop at the end of the working day, leaving a single double socket power source for testing electric clocks over night. This ensures that all of that outdated equipment is fully shut down whenever you are not on site.

Hazardous materials come into play on a daily basis when working with clocks, from the propane source for your blowtorch down to the ammonia in the cleaning fluids. Adequate ventilation is a must, if possible I do recommend locating cleaning equipment and gas supplies outside of the building in a shed or converted outhouse, but if this is not possible, an electric ventilation fan, ducted to the outside of the building is wise, although a large open window will suffice if security is not impaired.

Of course it is also wise to take full advantage of personal protective equipment to avoid personal injury. Safety glasses when working the lathe or grinder, Leather gloves when using the polishing mop or mainspring winder, industrial rubber gloves when working with cleaning fluids etc. and of course regular cleaning routines, and fire extinguishers close to hand.

Disposal of hazardous waste should be taken care of by a certified professional. It is a good idea to keep old cleaning fluids well sealed in a garden storage container until several gallons have been collected, when a collection can be made at a lower cost than several intermediate collections.

Chapter 4

The Timepiece

INTRODUCTORY NOTE ON CLOCK THEORY

Before we start looking at the intricacies of the different types of clock mechanism it is useful to stand back and consider the theory involved.

Horology is a highly respected science to which history's greatest minds have made their world-changing contributions – such names as Isaac Newton, Robert Hooke and Galileo Galilei, not to mention the specialist horologists such as Tompion, Graham, Breguet and the late George Daniels. With hundreds of years of advances by the world's greatest minds, it is little wonder that the following chapters, although only skimming the surface of the subject, are quite lengthy.

This chapter and the two following are not intended to be a definitive theory of horology. They cover what I believe to be the minimum knowledge of the many subjects in the competent clock repairer's memory. Each subject can be studied in further detail, but the information provided is a solid starting point to help you understand and make sensible decisions during any repair.

To make reliable repairs, it is important that you understand the interaction of the components, what they do and the materials they are made of. Historical notes are included where practical lessons can be learnt, but they are not emphasized because they are not necessary to know from an engineering point of view.

It is important to note that there are no set rules in horology beyond the laws of physics, and because of this there are thousands of designs out there. I will try to generalize and cover the most common designs, but there will always be exceptions which seem not to comply. These exceptions can be accommodated once you have the knowledge contained within this book.

THE TIMEPIECE

A timepiece is a basic clock, which shows the time on a dial, but does not strike the hour or quarters and has no other complications such as date-work, moon phase, alarm, etc. It is made up of various sub-assemblies, all of which con-

Pair of clock plates with screwed brass pillars.

Example of a bolted steel pillar and prickled clock plates.

tribute towards moving the hands to show time. The sub-assemblies are the power source, the gear train, the escapement, the motion works and the oscillator (pendulum or balance). This is the simplest form of clock and the type with which we are to begin.

Timepieces come in many forms, with different escapements, motion works designs, pillars, etc., all of which will be covered later as we work step by step to ensure that you understand why a timepiece is designed as it is, and how it works.

Clock movements are built upon two (or more) brass plates a few millimetres thick. The thickness varies from clock to clock; thick plates are a sign of quality, but some eminent makers such as the Knibbs were known for using thinner than average plates with good results. The plates replicate each other in shape and size. One is the front plate, the other the back plate.

Brass was traditionally manufactured by hammering and scraping it flat. This process, which creates unequal stresses within the material, leads to stress cracks years down the line. The plates are hardened to help reduce the likelihood of distortion and wear. More modern clocks have their plates hardened by rolling the sheet brass under high pressure to reduce it to the required thickness (this results in more even stresses within the material), while some plates have a series of small evenly spaced dents pricked into the surface, further increasing the hardness of the brass (it is a work-hardening material, which we will cover later) without decreasing its thickness.

Clock plates are separated by brass or steel pillars, which are normally riveted or bolted to the back plate. The pillars need to be sturdy and square-shouldered, and should fit snugly and perpendicular to the plates: if they are loose then the plates may shift or sag, causing problems with the sub-assemblies which lie between them. There are usually four pillars, one for each corner, but in better-quality clocks a fifth is often added near the power source.

There are several designs of pillar: on older clocks you will find them riveted to the back plate, passing up to a shoulder through holes in the front plate, at which point they are cross-drilled and held in place by a tapered pin. As

STRESS CORROSION CRACKING

This is the sudden cracking of stressed, vulnerable materials (such as the hammered brass of old clocks) when they are subjected to a corrosive environment (such as ammoniated cleaning fluids). The cracking of brass in ammonia is also known as season cracking.

Microscopic cracks gradually appear due to corrosives in the atmosphere, and may propagate suddenly when immersed in clock-cleaning fluids. Most commercial clock-cleaning fluids contain ammonia because it brightens brass beautifully but I have yet to see direct evidence of the adverse effects in this mild concentration of ammonia. A lot of debate remains around the use of ammonia within horological circles because the risk is real, so you should bear it in mind when dealing with clocks of great age or value, and consider an alternative cleaning method in some cases.

I have tested the effects of strong ammonia on brass by soaking various scrap components in 33 per cent ammonia (far stronger than the ammonia in cleaning fluids) and have seen a definite negative effect. The rear plate of a carriage clock I tested showed a corroded surface after one week, but no further deterioration over a further three months.

Ammonia corrosion of a brass carriage clock plate left to soak for three months.

The pillars of this clock are riveted to the back plate.

French clock with its pillars riveted to the front plate; taper pins hold the back plate in place.

Cut-away of a cross-drilled pillar demonstrating the squareness of the shoulder.

TAPER PINS

The tapered pin is commonly used in clocks, and it is useful for the repairer to hold a good supply. They are commonly manufactured in sizes one to twelve, plus 'universal' and 'dial' sizes.

To make your own pins, hold a piece of brass or steel wire in a pin chuck, rest it on a wooden block and spin it with one hand while you file across it. Hold the chuck at an angle to the file and it will soon take a taper. The angle of taper should mimic the taper of a broach.

It is considered bad practice by some to use steel pins in a steel hole or brass pins in a brass hole, while others consider it bad practice to use brass pins at all. You should make your own mind up on this as there are arguments for both views.

Like metals bind over time, due to oxidation and fretting (microscopic rubbing of the metals, causing the contacting surfaces to bind). Dissimilar metals bind due to galvanic corrosion; the two metals act as anode and cathode just like in a battery, with the cathode attracting electrons from, and corroding, the anode (usually the steel pin). In clocks, humidity in the air can be enough to start this reaction.

For a strong, lasting grip, it is best that one material yields to the other, using the material's elasticity to hold the pin in place. Brass is a more ductile material than steel, so will yield to either a steel pin or a steel hole.

As you can see the decision is tough, and some people feel strongly about their view. I personally think neither is right or wrong and I opt for using steel pins throughout. I have never seen this cause a problem, and it is a simple enough matter to drill out pins which do bind.

Filing taper pins by hand on a wooden block.

---

time went on it became common to replace the rivet and cross pin with screws or nuts, which helped to improve rigidity and reduce manufacturing costs.

THE POWER SOURCE

The power source is at the bottom of the movement between the plates and, for the purpose of this book, is either a spring or a weight. In the earliest clocks in the towers of cathedrals (Salisbury Cathedral's clock is the oldest in the UK, dating to the 1380s) weights were used to power the movement, dropping slowly down weight chutes as the escapement allowed the gradual release of their energy. So as it is the natural progression, the weight as the power source will be covered first.

Weight-Driven Clocks

Weights provide a consistent driving force because their mass and gravity are constant. This makes them an excellent power source for a clock because fluctuations in power can cause inconsistencies in rate (ticks per hour).

Different types of clock use different size weights to meet their power needs: finer clocks use a significantly lighter weight than the average longcase. However, from a basic theory point of view, a weight is a weight, provided you use a sensible amount or a 'standard' clock weight for that type.

Lines, Chains and Ropes

Weights hang below the movement from a line, chain or rope. On occasions where space is a concern the line is directed upwards and over

WEIGHT SIZES

The following information is a rough guide; standards vary from maker to maker and by country of origin. If you are unsure, start light and add weight, provided the movement is in good order. Many clocks do vary from these figures but these make a good starting point if the original weights are missing:

- thirty-hour longcase – 8lb (3.6kg)
- eight-day longcase – 10lb, 12lb or 14lb (4.5kg, 5.5kg or 6.5kg)
- Vienna regulator – 2¾lb (1.25kg)
- Zaandam clock – 2¼lb (approx.) (1kg)
- cuckoo clock – range from ½lb to 4½lb (275g to 2kg); weights should match side to side
- modern three-train longcase – range from 4lb to 10lb (2kg to 4.5kg) and the chime train weight is always the heaviest.

Three forms of line for weight-driven clocks: traditional gut-line, chain and rope.

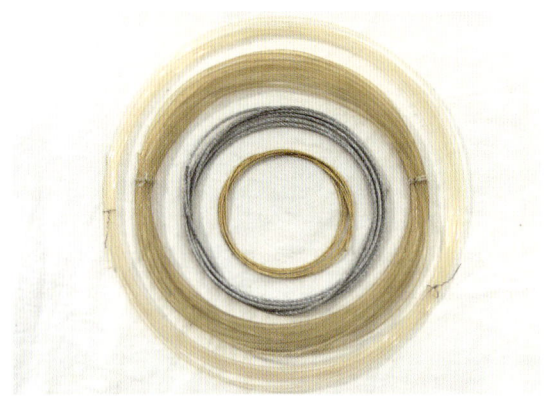

Lines for suspending weights: (from the outside) Perlon, traditional gut-line, braided steel and bronze.

Fracturing of steel lines when repeatedly bent through 90 degrees.

a pulley wheel. Thirty-hour longcase clocks use a length of chain, or rope formed into an endless loop. Eight-day longcase clocks traditionally use gut-lines, made from twisted sheep gut. Metal lines made of steel or bronze wire have a weakness, in that they fatigue and break when bent repeatedly at a sharp angle, which is exactly what happens as the wire enters the barrel. As the clock runs down the line is straightened; as the clock is wound, the line is bent through 90 degrees. I avoid this type of line where possible in favour of traditional gut-line or its synthetic counterpart, Perlon. Traditional gut is an excellent, strong material but I have found it to be sometimes brittle. Perlon is a compromise but has its place on clocks where originality is not the prime objective.

The weight hangs, via a hook, from a pulley wheel, which rides along the line coming down from the movement and looping back up to the seat board, where it is either tied into a loop and hung from a hook, or passed through a hole and tied into a 'butterfly'.

Applying Force

Whichever line type is used, the force must drive the gears by turning an arbor which is attached to the greatwheel. In order to cause it to turn, the force must be applied at a distance from the centre on which the arbor rotates. The arbor therefore acts as a centre for the barrel, which acts as a lever. The diameter of the barrel defines the length of the lever. The line wraps around the barrel as the clock is wound and naturally pulls on the barrel in an attempt to unwind itself. In rope or chain clocks, the barrel is replaced by a toothed pulley or sprocket, and the excess line hangs down below the movement rather than wrapping around the barrel.

Chain sprocket of a modern longcase clock.

Spring-Driven Clocks

Springs provide a convenient power source for driving a clock; they are small and compact, meaning that a tall case is not needed to accommodate them, and they allow the clock to be easily transported. The invention of springs to drive clocks opened up the possibilities of producing a variety of different types of clock including traditional English bracket clocks, carriage clocks and, of course, watches.

Clock springs are made of carbon steel or (more recently) stainless steel in long thin ribbons. They are coiled into the barrel, the diameter of which once again defines the distance between the arbor centre and where the force

ENDLESS LOOP

The endless loop system used in thirty-hour longcase clocks was first put to use by a Dutch scientist, Christiaan Huygens, whose name is often associated with the system. It allows the use of a single weight to drive both the striking and timekeeping gear trains of the clock, as well as providing a system to continuously provide power to the time train during winding, known as maintaining power, which has a positive effect on timekeeping.

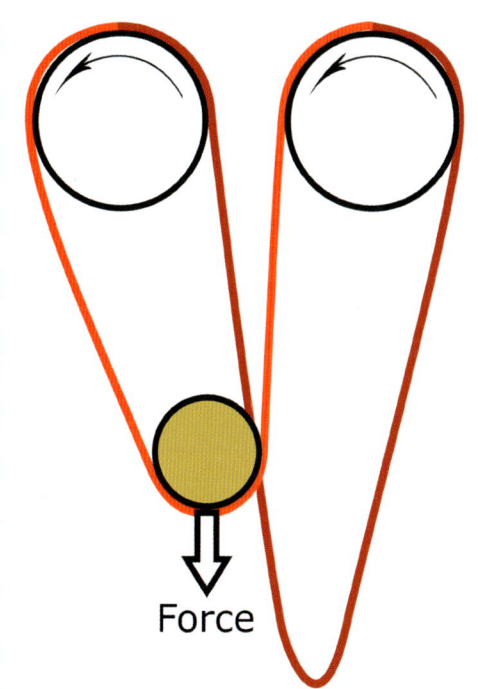

How the Huygens endless loop system provides power to both strike and time trains from a single weight.

CLOCKMAKER'S BUTTERFLY

Traditionally, longcase gut-lines would have been attached to the seat board by threading them through a drilled hole and tying a butterfly. To do this, create three or four loops about 2.5cm (1in) long in the end of the gut-line and fasten them with two half hitches, in opposing directions. The pull of the weight will keep the knot tight, while the loops will stop it passing through the hole.

Tying the clockmaker's butterfly: step 1, make a loop.

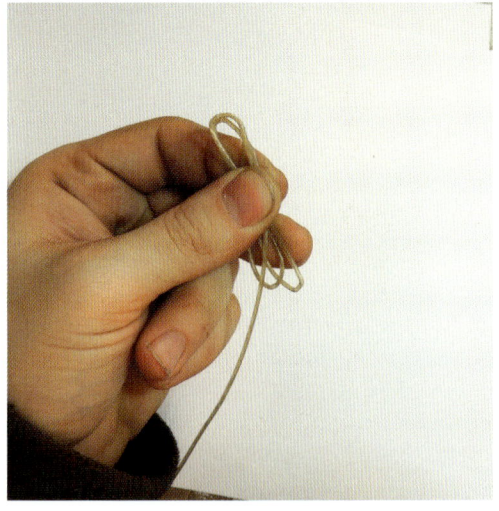

Tying the clockmaker's butterfly: step 2, make several loops.

Tying the clockmaker's butterfly: step 3, create a half-hitch knot.

Tying the clockmaker's butterfly: step 4, pass the loops through the knot.

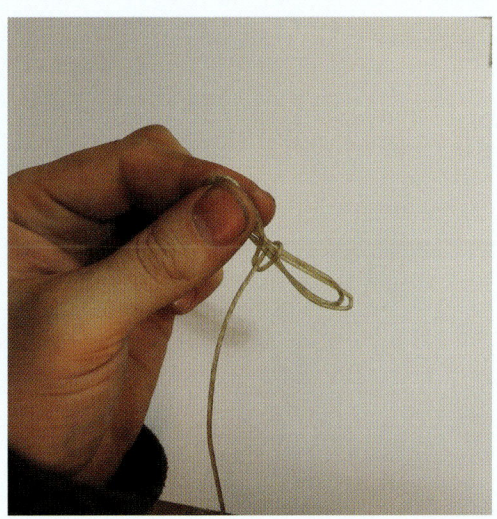

Tying the clockmaker's butterfly: step 5, tighten the knot around the loops.

Tying the clockmaker's butterfly: step 6, create a half-hitch knot on the other side of the butterfly.

Tying the clockmaker's butterfly: step 7, pass the loops through the half-hitch.

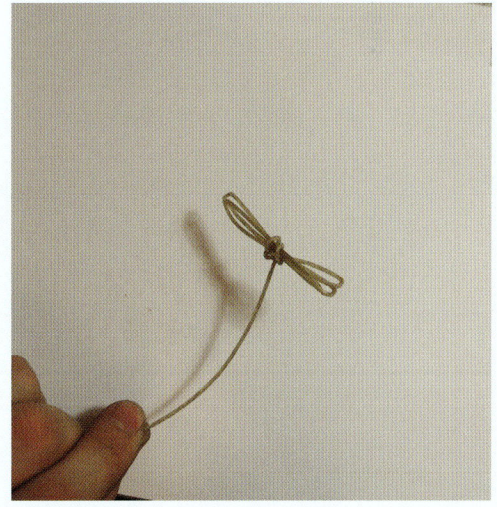

Tying the clockmaker's butterfly: step 8, the finished butterfly is self-tightening when under load.

38 *The Timepiece*

The long ribbon of steel that is a clock mainspring; when coiled it can be very powerful.

Clock mainspring loaded into the going barrel; note how the barrel arbor hook is well seated within the hooking eye of the spring.

is applied, or the length of the lever. The inside edge of the barrel is provided with a small hook, which grips a corresponding hooking eye on the outer edge of the spring. The barrel arbor has another hook to take the hooking eye on the inner edge of the spring. As the clock is wound the barrel arbor turns, taking the inner coils of the spring with it; the spring wraps around the arbor, pulling with great force at the hook on the barrel. The force is naturally trying to unwind the spring but to do so involves turning the entire barrel, to which the greatwheel is attached, driving the gear train.

Springs do not give the smooth and consistent power output that weights give. When fully

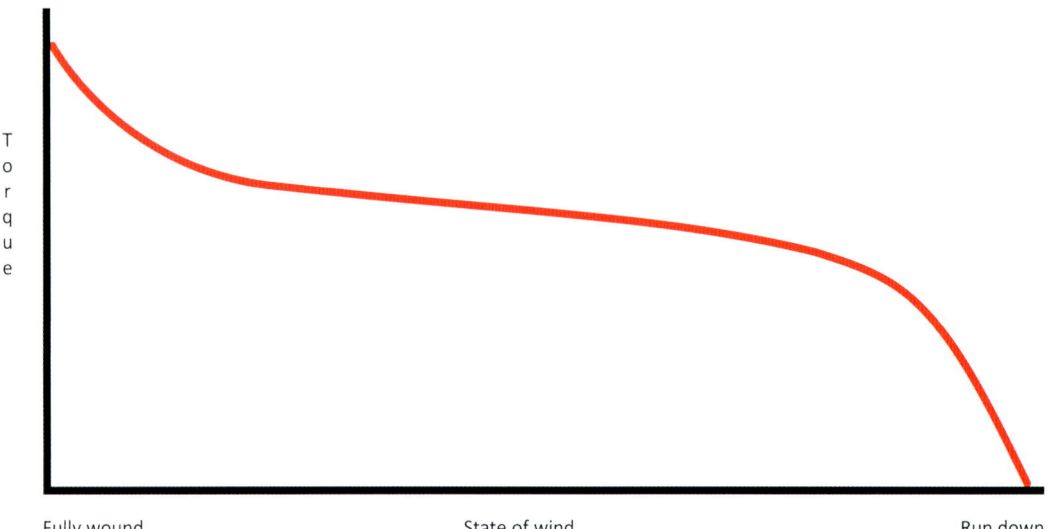

Power output of a mainspring as the clock goes from fully wound to fully down through the week.

wound, the power in the spring can be excessive, causing the clock to gain time, and when nearly run down power is lacking and the clock limps along losing time. Take the graph shown here, for example, which plots power output, or torque, on the Y axis against state of wind on the X axis. The shape of this graph is a rough estimate and does vary from spring to spring, dependent on material, thickness and temper and so on, but the general shape, depicting a power spike at fully wound and the sudden decline of power as the spring runs down, remains true.

Power Compensation

To improve the power output of the spring, there have been several types of compensation developed over the years. We will discuss the two most commonly encountered by the clockmaker: the fusee and Geneva stop work.

The Fusee: this a traditionally English component found in many older bracket clocks and is very common in English clocks. It is a conical brass casting with a spiral groove machined on its outer surface by a fusee cutting engine. At its centre, a steel arbor is pressed in, on which it pivots and on which the winding square is cut. A traditional gut-line or chain (smaller but similar in design to a bicycle chain) fits the groove and attaches to the fusee at its largest diameter via a hook or knot. The groove is cut with either a square or round bottom (square for chains and round for gut-line); the hooking method also varies depending on type of line used. The other end of this chain or line attaches to a barrel containing the spring. As the clock is wound, the chain wraps around the fusee, pulling hard at the barrel, rotating it and therefore winding the spring. The elastic nature of the spring is always pulling back and trying to unwind itself, driving the fusee, which carries the greatwheel and drives the clock.

When fully wound, the fusee comes to a physical stop, thanks to the fusee stop iron. This is an arm which, on contact with the fusee line or chain, is pushed into the path of a finger on the tip of the fusee cone. When they make contact, the fusee is physically no longer able to turn. The advantages of this piece are, firstly, a trimming of the extreme peak in power of a fully wound spring, by pre-

CARBON V. STAINLESS SPRINGS

Clock mainsprings were traditionally forged out of carbon steel, hardened and tempered to a dark blue. There are many advantages to this type of spring: the clocks we repair today were originally designed to work with the materials available at the time of manufacture and they can be tempered (softened) so that new hooking eyes can be made easily. Also, they tend to last for a significant number of years: I have seen many a 150-year-old French clock with its original springs.

Modern clock springs are being manufactured out of alloyed stainless steel. From an engineering point of view this does make sense: they do not rust, they are strong and do not tear easily, and they hold their elasticity for longer. But stainless steel springs are far too strong 'size for size', and they do not temper, so making new hooking eyes is not a simple task when springs need to be shortened or repaired.

Many clockmakers are of a similar opinion, that modern mainsprings made of stainless steel have no place in horology. I have seen many clocks damaged by use of a 'size for size' spring which is far too strong, and, where possible, I use replacement carbon steel springs, which are still available in the more common sizes (at the time of writing). If you have no choice but to use stainless steel springs, select a spring one 'size' smaller, either in width or thickness, to compensate.

Fusee, the arbor of which carries the winding square.

Fusee chain; note the different shapes of the hooks (the top hook is intended for the barrel, the lower hook for the fusee).

Fusee stop iron in action (shown here highlighted in red); the gut-line has pushed the stop-iron into the path of the fusee 'finger' as it was wound.

venting full winding, and a prevention of excess winding from damaging the fusee chain or line and possibly the clock winder's fingers.

The easiest explanation of how a fusee works is to imagine it as a series of levers and fulcrums (pivots). Consider trying to undo a nut: the lever is the spanner, the fulcrum is the nut itself, which acts as the pivot point. You apply a force, you pull as hard as you can but the nut will not budge. You know that by changing the length of the lever you can increase the amount of force seen at the fulcrum, so you slip a length of pipe over the spanner and pull again and the nut undoes with ease. Essentially this is what the fusee does: the distance from the arbor centre to its outer edge where the line is pulling is the lever, the force comes via the line or chain from the spring and the arbor is the pivot point or fulcrum. So when the spring is fully wound and pulling at full force, a small lever can be used, and as the spring gets weaker, the length of the lever increases at the same rate, resulting in an equal force seen at the fulcrum, or arbor, which in this case drives the greatwheel of the clock; an even force is the result.

It makes sense that, as the power curve varies from spring to spring, so should the perfect fusee cone. A change of spring in this case would require a change of fusee to match. In reality we do not do this; any fusee is better than none, and with temperature errors, circular error, tolerances, friction, etc., playing their part, there is no need for this level of accuracy, or to destroy the originality of the clock.

When the fusee is fully unwound (that is, with the line or chain fully on the spring barrel), there is still some power stored in the spring. It is important to know this for several reasons: firstly, for safety, never disassemble a clock until all of the power is out of the spring; secondly, it helps to keep tension on the line and keep it wound around the barrel; and finally, it trims the 'power curve' of its weakest portion allowing you to make better use of the spring. This is known as the 'set-up' power.

Stop Work: Another option when attempting to improve the power output of a spring is to clip the power curve of its extremes, and use only the central portion which is more even. To do this a simple physical stop can be put in place to keep you from fully winding the spring and to keep the clock from fully unwinding the spring. This is as far as the basic theory needs to go but a description of the pieces at work

Trimmed power curve of a clock with Geneva stop work.

will help to clarify things. The trimmed power curve can be seen in the graph shown here.

Geneva stop work is the variety most commonly found in quality clocks; there are other types of similar stop work but their operation is obvious once you understand the basics. The stop work consists of two pieces and a shoulder screw. One piece, which resembles a star, has a number of slots cut into it and a number of curved cut-outs the same diameter as its counterpart around its edge. The number of slots depends on how many turns the spring is intended to make. It fits via the shoulder screw onto the barrel cap. The other piece fits on the end of the barrel arbor via a square, so that it turns or holds steady in relation to the barrel arbor. It has a 'finger' which fits the slots in the star piece. They are arranged so that as the clock is wound, the barrel arbor, and therefore the finger, rotates. As it does so, it passes into one of the slots on the star piece, causing it to turn. The curved cut-outs make a space for the finger piece's shoulder to pass as it rotates; when fully wound, the star piece does not provide this cut-out and the shoulder of the finger piece butts against it. As the clock unwinds, the barrel rotates around the arbor and the finger piece turns the star piece in the opposite direction by a predefined number of turns until the star piece butts on the opposite shoulder and the spring will unwind no further. When setting up, it is the repairer's job to judge how much pre-wind to put on the spring before putting the stop work on in its locked position, so basic knowledge of the power curve is important.

Geneva stop work in its fully down position; note how the shoulder of the star piece is resting on the shoulder of the finger piece.

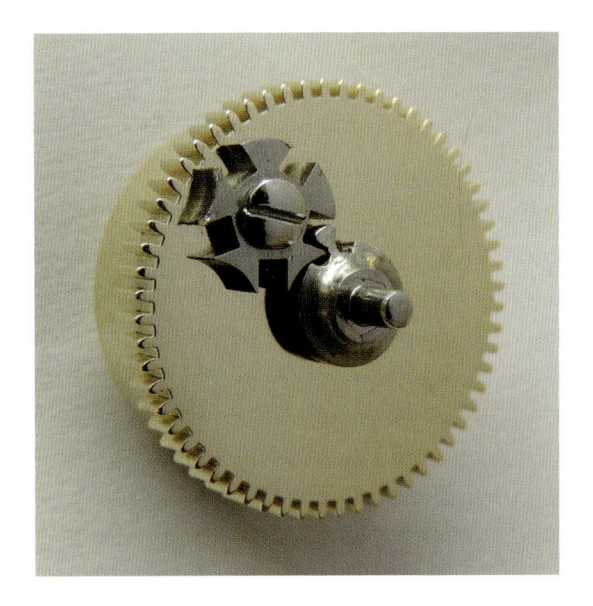

Geneva stop work when it is first wound; note how the shoulder of the finger piece causes the star piece to rotate.

Geneva stop work demonstrating how the hollow of the star piece provides clearance for the barrel arbor to turn.

Geneva stop work demonstrating that the finger piece will enter the star piece slot, causing it to rotate to the next hollow.

Geneva stop work in the fully wound position with no hollow to allow the star piece to rotate.

Click Work

By now you may be wondering what stops all the power from the spring or weight going not through the gear train but back through the barrel arbor, causing the winding square to rotate backwards violently when you let go of the key. The answer to this is click work.

The click work consists of a ratchet wheel, formed with triangular pointed teeth (think right-angle triangle, not equilateral), either left- or right-handed, and often a squared centre which fits at the base of the winding square.

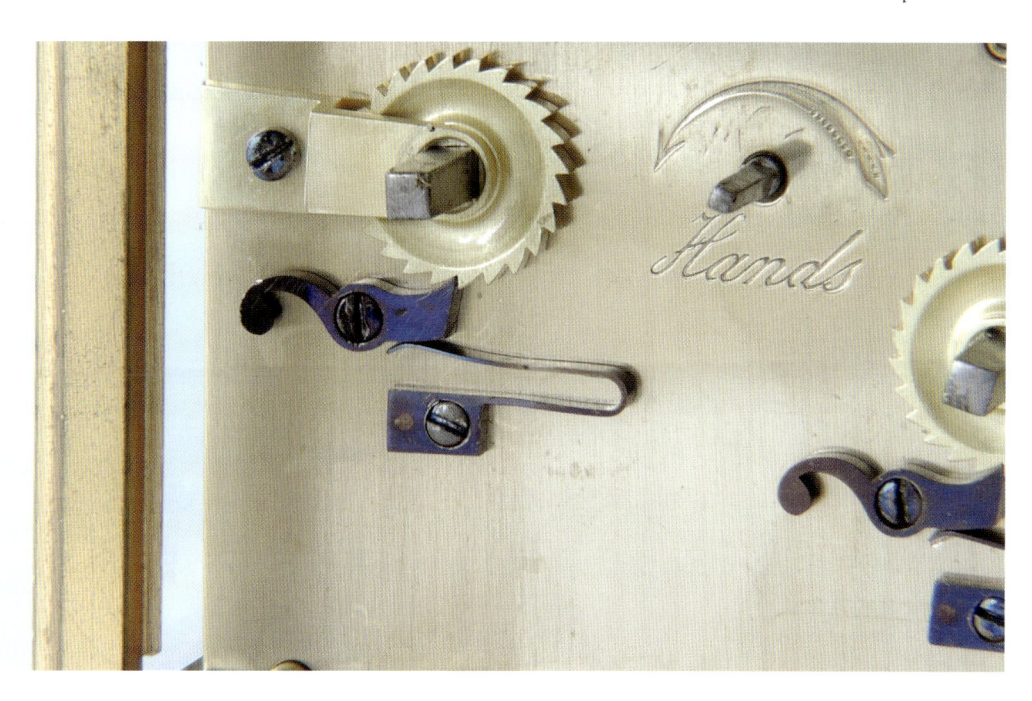

The most basic form of click work is used on even the best-quality carriage clocks.

The next piece is a click, which pivots freely on a shoulder screw and is shaped to fit the ratchet teeth perfectly when set at a tangent to the ratchet wheel. Then there is a click spring, used to keep the two in firm contact. As the clock is wound, the barrel arbor and therefore the ratchet wheel rotate in the direction which causes the click to ride over the directional teeth. When you release the winding key, the barrel arbor is forced backwards by the power in the spring, but the ratchet teeth are caught by the face of the click, redirecting the force to its pivot, the shoulder screw, and holding back the power of the spring, giving the spring no option but to power the gear train.

Click work comes in several forms; on longcase clocks the ratchet wheel is formed at one end of the barrel and the click and click spring are mounted on the greatwheel. On fusee clocks the ratchet wheel is on the rear of the fusee attached by screws; it is then covered by the greatwheel which carries the click and click spring. Fusee clocks have a second click and ratchet on the front plate used for setting up the fusee spring. However, this is not the same; the click is screwed solidly to the plate and must only move when being adjusted by the repairer.

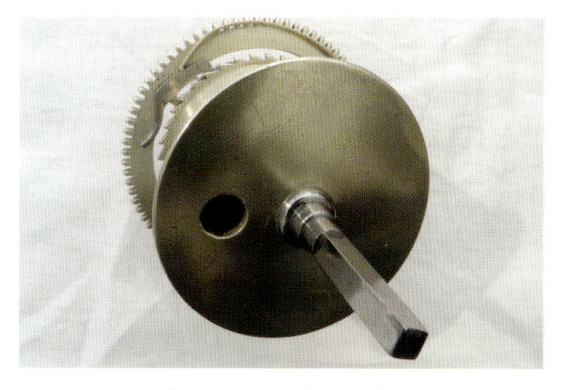

Longcase barrel showing its click and spring mounted to the greatwheel.

Fusee click work is mounted within the greatwheel assembly and cannot by seen during normal use.

GEAR THEORY

In horology there are wheels and there are pinions. Pinions are generally made of steel, have fewer than twenty leaves and are the driven gears, or followers. Wheels are cut from brass and are the drivers; they are riveted to the rear of a pinion or to a brass collet soldered onto the arbor.

Horological gear trains are geared up, not down; this means that the speed of rotation at the final gear is significantly faster than the first. The increase in speed reduces the torque available, meaning that wheels higher up the train can suffer from lack of power and increased sensitivity to friction. In other common gear trains such as in cars or electric motors, the train is geared down to slow the output but increase the torque into usable power.

Pinions

Pinions come in two forms, machined and lantern. Machined pinions are found in most higher-quality clocks, and are usually cut directly onto the arbor; the steel is hardened and tempered to decrease wear. The teeth of pinions are called leaves, and the shape of each leaf is important as it become distorted with wear. Lantern pinions are used in cheaper, mass-produced clocks, commonly in those from America and Germany. They resemble a bird cage and are made up of multiple pieces. Two brass rings or caps are pressed or soldered onto the arbor at a distance which defines the length of the pinion. A series of equally divided holes are drilled around these caps, into which the trundles are pressed. The trundles are made of polished and hardened round steel bar and act as the leaves of the pinion. Lantern pinions are highly repairable, unlike machined pinions. The type of pinion in use theoretically dictates the tooth shape for the meshing wheel.

Ratios

As already stated, horological trains are geared up, and this is a result of the larger wheels driving the smaller pinions. When a wheel and pinion in constant mesh with each other are caused to rotate, the smaller pinion will be forced to make more turns and to travel at a higher speed

Machined pinion.

Lantern pinion.

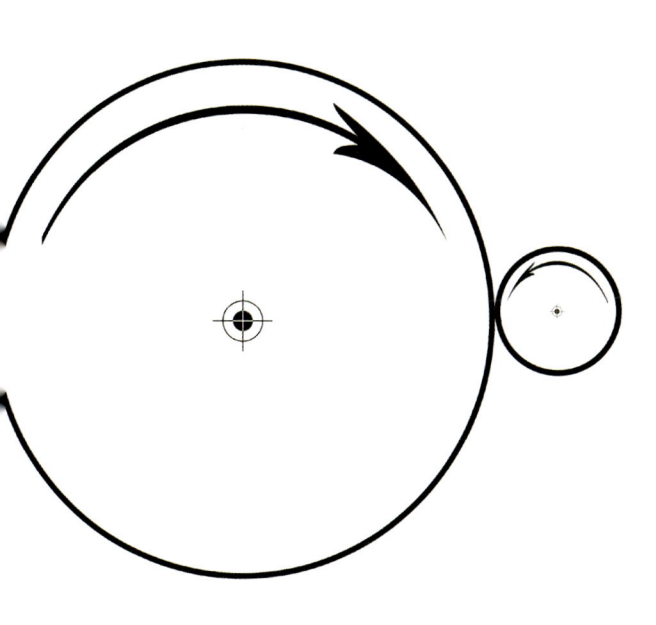

The smaller wheel is a quarter the size of the larger wheel; it will therefore rotate four times for every one rotation of its larger partner.

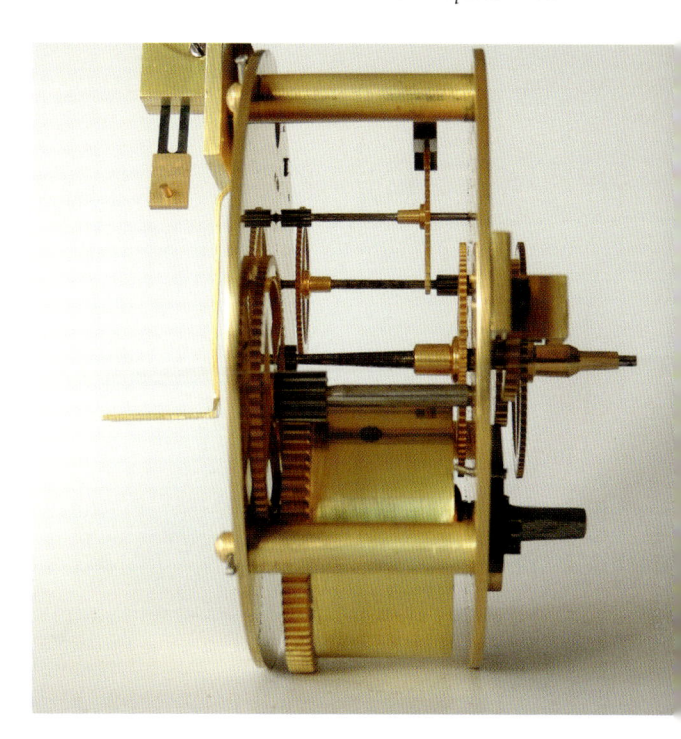

Side view of a standard timepiece movement; note how the wheels get smaller higher up the train.

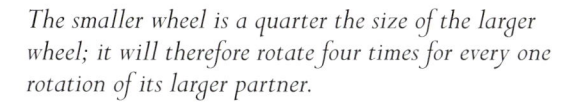

in order to keep up with the larger wheel. For example, a pinion of one quarter the diameter and number of teeth of a wheel will need to rotate four times for every one rotation of that wheel. This is a ratio of 4:1 (four to one) and this is the basis of the calculations used to design gear trains.

The power source drives the greatwheel; this is the first gear in the train, and in order to ensure that the clock runs for an entire week, this wheel only rotates once or twice a day depending on design. The greatwheel then drives the second arbor pinion. The second arbor is an intermediate arbor that is not always used, and very rarely in longcase clocks. The second wheel, which turns with its arbor, drives the centre pinion. The centre arbor is always designed to rotate once per hour, as it carries the minute hand of the clock. The centre wheel drives the third pinion, which again is an intermediate gear, designed to ensure that the final arbor, the escape arbor, rotates at the correct relative speed (once per minute if it holds the second hand).

Gear Depthing

Correct gear depthing is important if the clock is to run efficiently, or at all. A wheel and pinion are correctly depthed when their pitch circles are 'touching'. Imagine two smooth disks rotating side by side with friction alone causing them to turn together; one of these is the driver, and the other is the follower. They are essentially gears without teeth. Adding teeth to these will stop them slipping when knocked or under power fluctuations, but to avoid altering the ratio (which is dependent on their relative sizes), you cannot just add teeth to the outside of one wheel because this would make it bigger, nor can you cut hollows into the other without making it smaller, which would alter the pitch circles and ratios. Instead material is both added and removed from each wheel, in the right proportions to keep the pitch circle correct for the ratio required. The addendum is the protrusion which stands proud of the pitch circle in order to protrude into the pitch circle of the opposing wheel to 'grip' it. To accept the addendum,

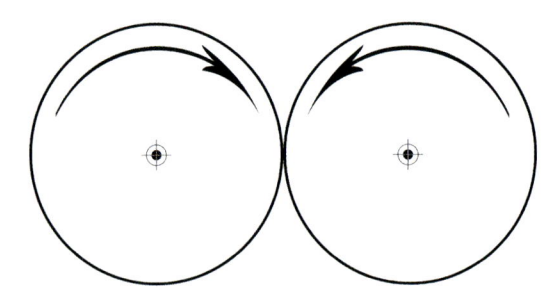

Two smooth wheels in frictional contact will rotate in opposing directions at a speed dependent on their relative size.

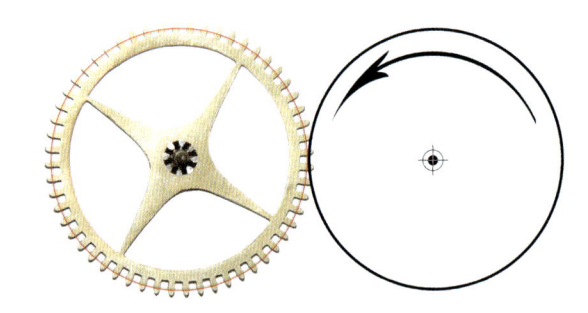

Addendum and dedendum added to one wheel; the pitch circle shown in red is now theoretical.

a trough is cut into the opposing wheel, the dedendum. Both wheels have addendums and dedendums in order to accommodate each other. An addendum and dedendum make up one tooth and one tooth gap, and are repeated around the entire diameter of both wheels. The pitch circle is now a theoretical diameter and cannot be seen.

The shape of the teeth is designed to reduce friction. A perfectly made gear will 'roll' each tooth through its corresponding leaf thanks to this design; however, there are manufacturing tolerances, working freedom, etc., to take into account so, in reality, no wheel or pinion meets this level of perfection and a certain amount of freedom is designed into the tooth form to allow for it. In reality, we have sliding or engaging friction to contend with, where the teeth of the wheel slide over the pinion leaves rather than roll. This property is made worse on low count (leaf number) pinions, due to the point of contact happening before the line of centres, and is the reason higher counts are used in quality clocks; higher count pinions are weaker at the roots so a compromise must always be made. Correct depthing keeps engaging friction to a minimum, but disengaging friction is unavoidable. As pivots and pinions wear, both the depthing and tooth forms become distorted, causing increased friction and in extreme cases butting of the teeth and leaves.

The best way to learn about gear depthing is to study the images and then study a real-life example. If possible, try them in a depthing tool where you can change the gear depthing to get a feel for how a correctly depthed gear runs. When correct, the action will be smooth and quiet; incorrect depthing feels bumpy and stiff and is very noisy when spun. There is a certain amount of freedom necessary between the

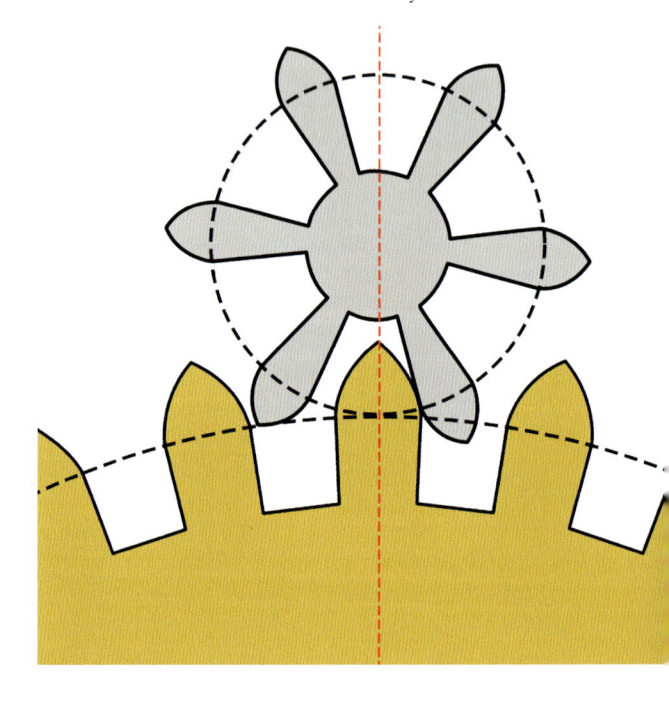

Wheel and pinion pair showing their pitch circles and line of centres.

LINE OF CENTRES

The line of centres is an imaginary line connecting two meshing gears between their centres of rotation (pivots).

The perfect wheel and pinion will be a pair in which the initial point of contact between tooth and leaf happens on this line. In this model, the gears will roll smoothly through each other, with no engaging friction and no 'play'.

In the perfect model, any slight manufacturing inaccuracy, particles of dust or even the freedom in the pivots necessary to allow movement would be enough to cause the teeth to butt and to cause the clock to stop. Therefore a certain amount of freedom is designed into the cutters. This freedom causes the initial point of contact to happen away from the line of centres.

Pinions with low leaf counts (take six as a common example) cause problems because the initial point of contact occurs well before the line of centres, causing engaging friction, where the tooth slides across the leaf surface.

Higher leaf counts move this initial point of contact towards the line of centres and reduce engaging friction.

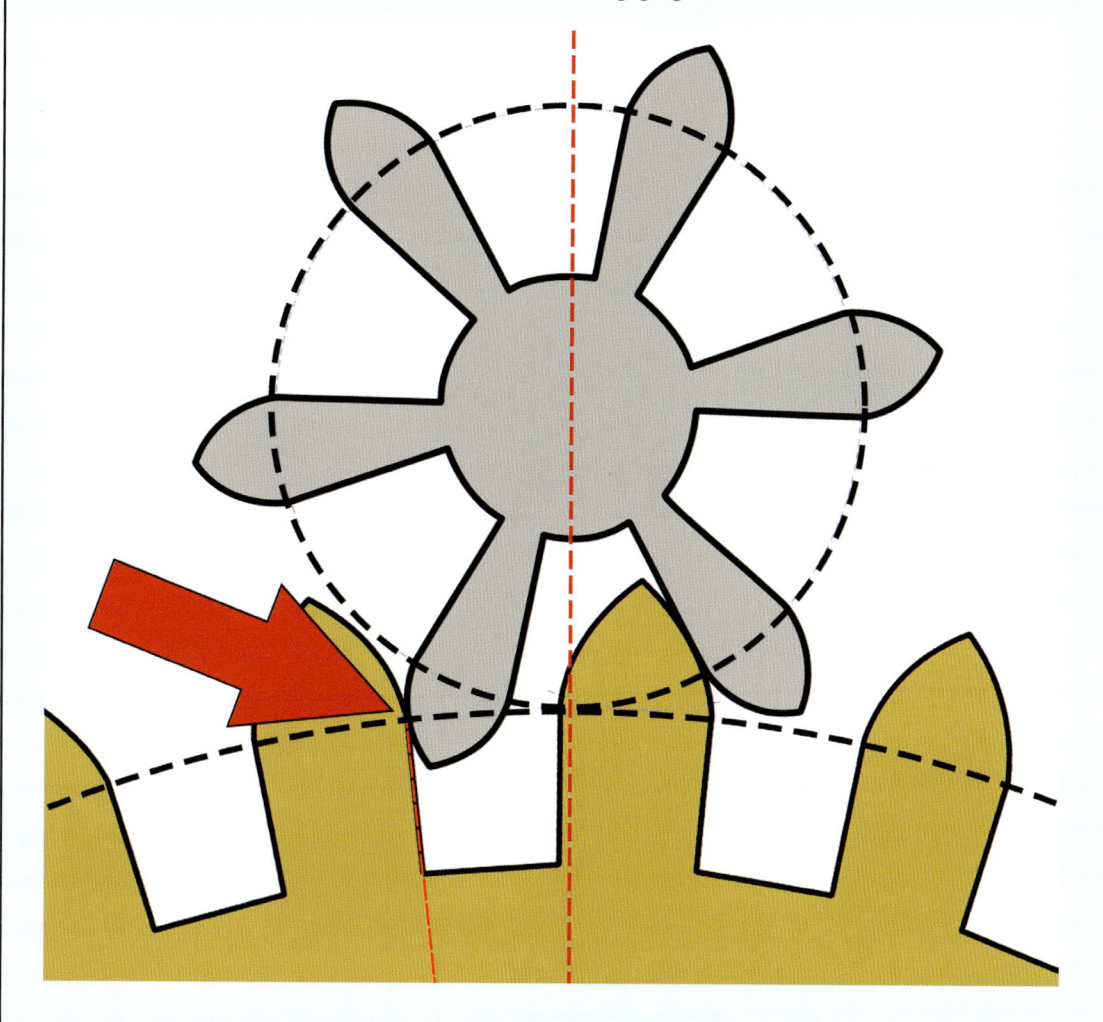

The initial point of contact shown here well before the line of centres.

CONTRATE GEARING

In carriage clocks it is necessary to turn the gear train through 90 degrees to drive the escape pinion mounted in the platform on top of the movement. To do this, contrate gearing is used. The contrate wheel has its teeth cut to face forwards, rather than outwards. This allows the gearing to act as usual (near enough) but with the escape arbor at 90 degrees.

There is the problem of overcoming thrust, which pushes the contrate wheel out and away from the pinion. To correct this a hardened screw is fitted to the back plate on which the contrate rear pivot rests, with its end-shake therefore adjustable. The depthing of the contrate wheel to escape pinion is adjusted by moving the entire platform back and forth.

The thickness of the contrate teeth, or thickness of the 'rim', has an effect on depthing. As the contrate teeth enter the pinion, the bottom edge of the tooth contacts the pinion leaf; the tooth then changes angle at it rolls through the pinion, finishing with the top edge of the tooth, effectively making the tooth 'appear' thicker to the pinion than it really is. Occasionally repairers thin the teeth of troublesome gearing to reduce this effect, often introducing more problems by bending and weakening the contrate teeth.

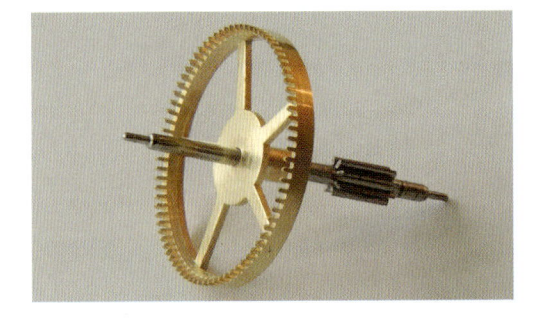

Carriage clock contrate wheel with its teeth orientated to change the direction of drive through 90 degrees.

wheel and pinion when correctly depthed, even more so in low count pinions, a characteristic which is in their favour as it makes them less sensitive to dirt and wear; you will eventually learn to judge this by sight and feel.

Motion Works

The motion works is a set of three gears: the cannon pinion, the minute wheel and the hour wheel. Their job is to change the once-per-hour rotation of the centre arbor, which extends through the front plate and out beyond the dial, through a 12:1 reduction, into a once-every-twelve-hours rotation. This leaves you with the centre arbor turning once an hour, taking with it, through a friction drive, the cannon pinion. The cannon pinion drives the minute wheel, which turns at the same rate, but in the opposite direction (anti-clockwise) which through its own pinion drives the hour wheel (the only common use of a pinion as the driver) at a 12:1 ratio in a clockwise direction. You now have

Typical French motion works showing the direction of rotation for each wheel; note how the minute and hour wheels rotate together but at different speeds.

two gears turning in a clockwise direction, one turning at a rate of once per hour and one at once every twelve hours, to which you can attach the minute and hour hands. The motion works is arranged in such a way that the hour wheel rides directly over the cannon pinion, allowing both of the hands to rotate on the same centre.

Friction Drive

In order to be able to set the hands, there must be a friction fit of the motion works onto the centre arbor. If the motion works were directly attached to the centre arbor, then setting the hands would force the entire time train to turn, causing damage.

The friction fit is largely one of two types: either the cannon pinion itself is made an interference fit on the centre arbor, which directly causes the friction, or a hammered brass spring is placed under the cannon pinion, resting on a shoulder on the centre arbor. As the minute hand is fitted, it presses the cannon pinion against this spring, producing friction between the two. The friction is just enough that the centre arbor drives the motion works, but slips under hand setting. More modern, mass-produced clocks tend to have the centre wheel itself slip on the arbor, with friction provided by a spring behind the centre wheel, and the whole arbor turning when the hands are set.

ESCAPEMENTS

The escapement produces the distinctive tick of antique clocks. There are hundreds of designs patented, but there are only a few which became popular and are in common use. We will cover five of the most commonly encountered escapements: the recoil anchor escapement, the deadbeat anchor escapement, the brocot escapement, the cylinder platform escapement and the lever platform escapement.

As you will see, the main functions of the escapement are to control the speed of the going train (or to control the release of power from the power source), and to keep the pendulum or balance swinging by periodically providing impulse. Understanding this will help you understand the basic working of any escapement which you may encounter.

Friction-fit cannon pinion of a French movement.

Friction spring cannon pinion of an English movement.

Centre wheel friction setting found in most modern clock movements.

The Crutch

The crutch is the component which creates the relationship between the swinging pendulum and the pallet arbor of the escapement. It is necessary to understand the role of the crutch in order to understand how pendulum escapements work, which is why I have chosen to cover its use in this first. Altering the crutch by bending (or by slipping of the friction fitting) alters the relationship between pendulum and escapement and is a technique used to set many clocks in beat. There are multiple designs, but a general explanation is as follows.

The crutch extends perpendicular to the pallet arbor (usually downwards, but there are exceptions) which carries the anchor of the escapement; the pallet arbor is over-long and extends past the back plate where its rear pivot is supported in the back cock. The back cock allows the crutch to hang behind the rear plate and parallel to it; it also acts as a base for the pendulum to hang from. The pendulum hangs parallel to the crutch; the crutch then turns outward toward the pendulum or has an attached piece at a 90-degree angle, and from this point outward is referred to as the crutch foot, which comes in several forms. The crutch foot, which is parallel to the pallet arbor, 'interferes' with the pendulum rod so that they pivot together while still allowing the pendulum to swing freely. As the pendulum swings, it takes the crutch with it, causing the pallet arbor to rotate in alternating directions, activating the escapement.

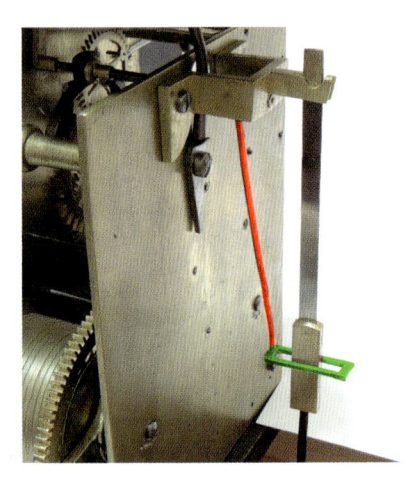

Crutch rod of a longcase clock (shown here in red) and the crutch foot (in green).

Recoil anchor escapement of a fusee movement.

The Recoil Anchor Escapement

The recoil anchor escapement is possibly the most commonly encountered of all escapements. It is found in most English longcase clocks as well as in its various modified forms in clocks from all origins. As they vary only slightly in design, I will focus on the type commonly found in English longcase and bracket clocks.

The escapement is made up of several parts: the escape wheel, the anchor, the pallet arbor (which carries the anchor and crutch) and the back cock. The escape wheel is attached concentrically to the escape arbor which is driven

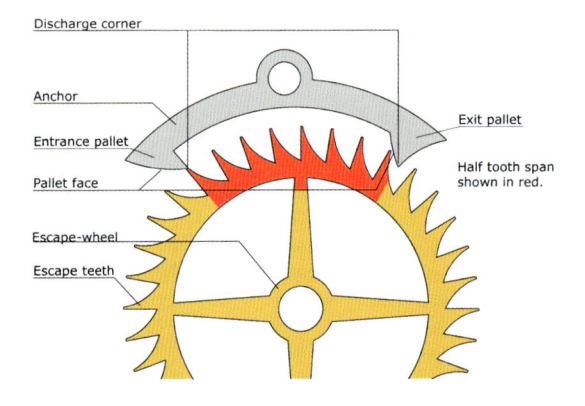

Diagram of the recoil anchor escapement.

by the third wheel. It therefore has constant drive from the power source, causing it to rotate. The anchor controls the speed of rotation of the escape wheel by catching and releasing it one tooth at a time.

The anchor has two pallets on which the escape teeth slide during both recoil and impulse. They are made of hardened steel to reduce wear, but do become pitted with time. Each pallet has both a face and a discharge corner. The pallets are always spaced a half-number of escape teeth apart to place a pallet in a tooth gap while the opposite pallet is holding up a tooth and also to ensure that one pallet is always within the escape wheel diameter so that the escapement cannot 'run through'.

The escapement cycle is made up of four parts: drop, lock, recoil and impulse; and this cycle repeats continuously. We will start with the entry pallet (the first pallet a tooth encounters when 'entering' the anchor) about to release a tooth of the escape wheel.

Imagine watching the escapement from the front, with the pendulum swinging to the left, and the escape wheel with a tooth nearing the discharge corner of the entrance pallet. The anchor will rotate clockwise about its pivot, raising the entrance pallet and unlocking the escape wheel, which will then begin to drop.

Drop

Drop is the distance for which the escape wheel is in free fall before being caught by a pallet. Drop is wasted power and should be kept to a minimum, but it is important that it is large enough to provide ample clearance between the wheel and anchor so that there is no danger of collision; poorly made escape wheels require larger drops to be safe. The wheel may be mounted eccentrically: the teeth may be unevenly spaced, tall or short, etc.

Lock

Eventually a tooth further around the escape wheel is arrested by the exit pallet of the anchor, seven and a half teeth away.

Lock, or the locking point, is the point at which the escape tooth lands on the pallet after dropping. The depth at which the escapement locks is important: if it is too close to the discharge corner, then the escapement may not lock safely. For the recoil escapement, deep locking is good as it reduces recoil and drop. If the escapement locks too deeply, then the anchor will butt the top of an escape tooth on its return swing; finding the correct depth can be challenging on poorly divided escape wheels or on those with teeth of various lengths.

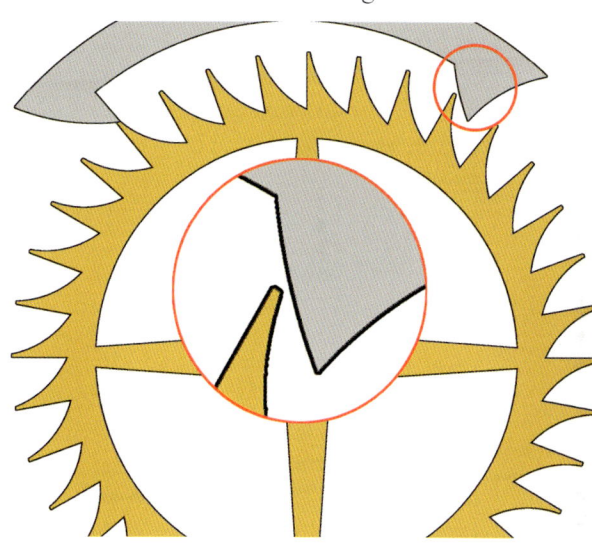

Recoil escapement demonstrating the amount of drop onto the exit pallet.

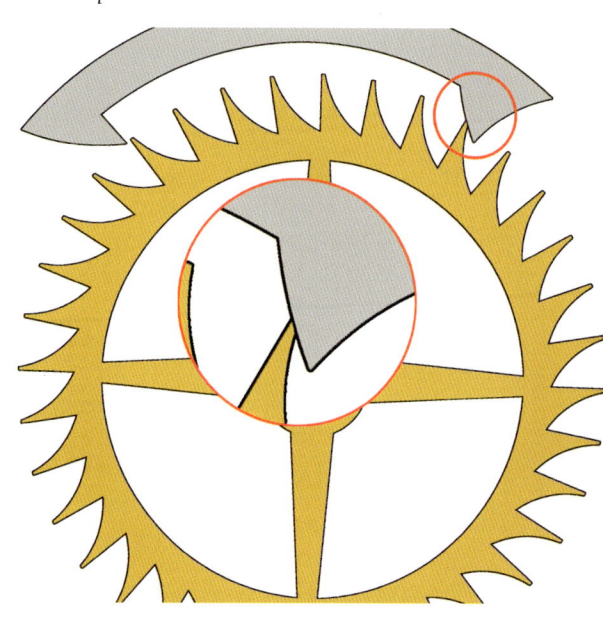

Recoil escapement having just locked the escape tooth onto the exit pallet.

Recoil

As the pendulum swings to the right, with an escape tooth locked on the entrance pallet, the pallet is forced to move deeper into the escape wheel, forcing the wheel to rotate in reverse against the force of the train. This is called recoil, which is wasted power as it steals momentum from the pendulum. It is detrimental to timekeeping and is the major issue with these escapements. The recoil anchor escapement is a huge step forward from its predecessor, the verge, as it is far less sensitive to power fluctuations and requires a smaller degree of swing from the pendulum, reducing circular error.

Impulse

As the pendulum reaches the highest point of its swing, it is slowed by gravity and the friction of recoil, and reverses direction. At this point the pendulum begins to free fall, allowing the escape wheel to rotate and the tooth to slide back along the pallet face. As it does this, power from the weight or spring pushes the tooth along the pallet face and forces the anchor out of its way; this energy is passed via the crutch as a small impulse to the pendulum to keep it swinging. This pushing on the pallet face lasts well into the uphill climb of the pendulum before the discharge corner is reached. This process is called impulse. The pallet faces should be polished to reduce friction.

As the pendulum continues its swing to the right, and the exit pallet rises out of the path of the escape tooth, the tooth reaches the discharge corner and drops off the anchor; the escape wheel drops freely until a tooth is arrested by the entrance pallet, where the process is repeated.

The four steps, drop, lock, recoil and impulse, are important to understand and are best learnt by studying an escapement and taking note of the action. A good understanding makes tackling uncommon escapements a far simpler task. As drop, lock, recoil and impulse generally do not differ in description, I will refer to them by name from here on.

A variation of the recoil anchor escapement is the strip pallet variety found in American and German clocks; this is made from a strip of bent and hardened steel. The geometry, however, is

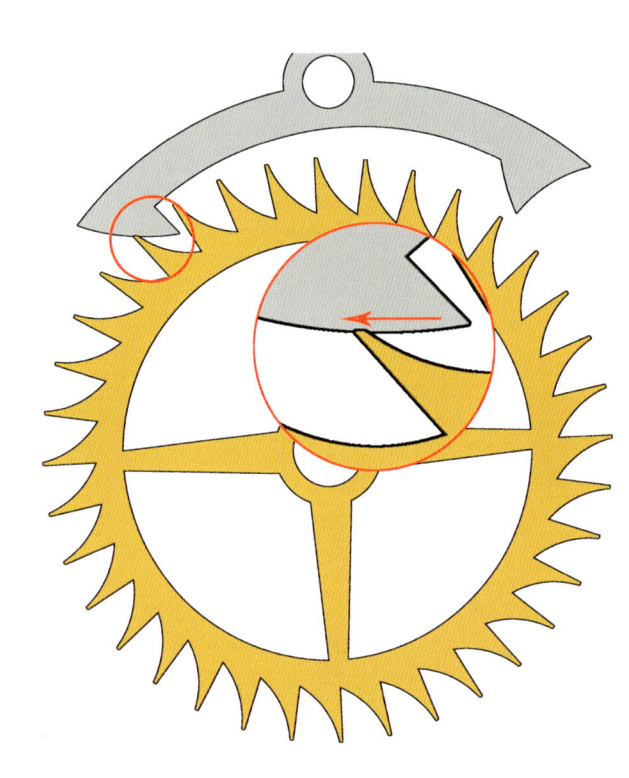

Recoil escapement demonstrating recoil on the entrance pallet.

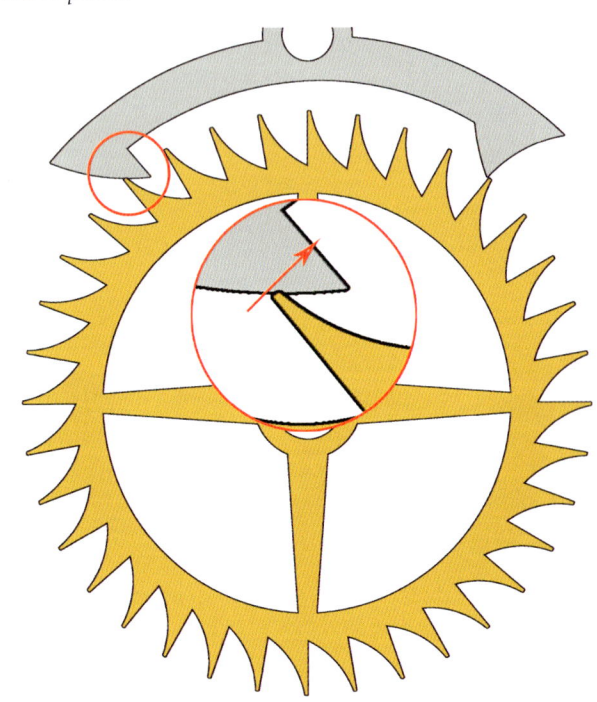

Recoil escapement demonstrating impulse on the entrance pallet.

Solid anchor and strip anchor compared; the pallet geometry remains similar.

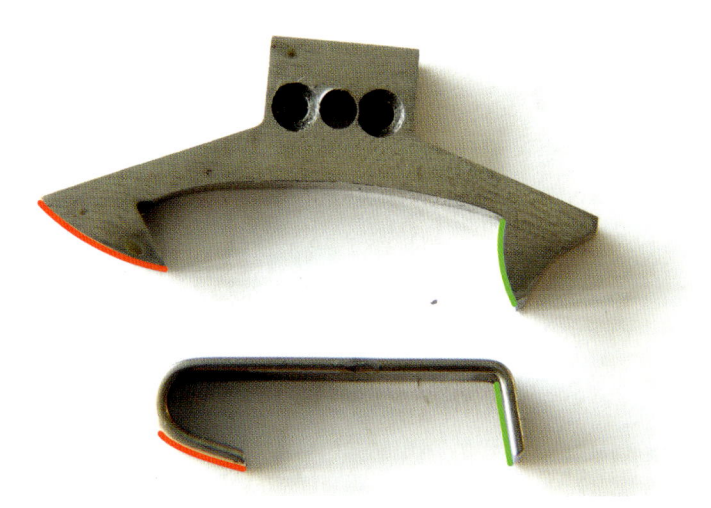

the same. Often they are found mounted on the front plate of American movements; the escape arbor protrudes through the plate with the escape wheel on the outside, pivoted in a cock. The strip pallet anchor is then (often) beneath the escape wheel, pivoted on a pin.

The Deadbeat Anchor Escapement

The deadbeat anchor escapement (invented around 1720 and known as the Graham deadbeat escapement, after its inventor George Graham) is an improvement of the recoil anchor escapement. The improvement comes largely from the abolition of recoil which is achieved by redesign of the pallet faces. The single pallet face and discharge corner of the anchor become two faces – the locking or dead face and the impulse face – and the two corners become the locking corner and the discharge corner. The deadbeat escapement did not replace the recoil escapement in most types of clock. The escapement has various weaknesses, so it is not suitable for all clocks. Its finer proportions are great for

Deadbeat escapement of a good-quality Dent regulator.

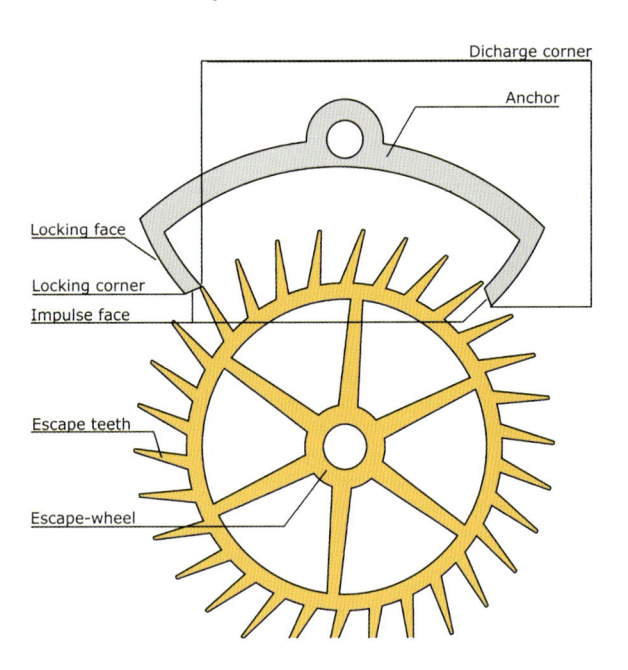

Diagram of the deadbeat escapement.

A jewelled deadbeat escapement reduces friction and increases the accuracy and reliability of the clock.

accuracy, but make the escapement more sensitive to beat and more easily damaged, especially in clocks which are likely to be moved around.

The escape wheel is made from hard brass, sometimes with the rear faces of the teeth cut away to form a passing hollow for the pallets, and the tooth tips are much finer than the recoil escapement; they are not subject to the forces of recoil and need not be so strong, therefore the advantages of reducing friction can be fully realized.

The locking face is curved and is concentric to the pallet arbor so that as the anchor rotates with the pendulum its face does not move relative to the escape tooth. As the escape tooth is resting on the anchor, it will remain motionless until the anchor presents the locking corner to the tooth, at which point impulse will begin.

The impulse face should be at the correct angle in order to be at its most efficient. The angle, to the beginner, should allow for visible impulse: that is, the escape wheel should rotate as it passes along the impulse face by an equal degree on both entrance and exit pallets.

A correctly adjusted deadbeat escapement will lock safely on the locking face after drop; if the escapement is locking on the impulse face, then it will recoil, causing damage to the escape teeth.

After locking, the tooth will rest on the locking face during the part of the process which was previously called recoil; here we will call it rest. The resting period of the deadbeat escapement is small; friction between the tooth and pallet face is detrimental but negligible in most cases.

On some finer clocks, the pallets are cut from jewels: ruby, sapphire or one of the synthetic materials used for jewel-holes, end stones, lever pallets, etc. These materials are brittle and easily damaged; however, they are very hard-wearing and, when coupled with jewelled pivot holes, can produce a very accurate and long-lasting escapement.

As the pendulum reverses direction, the escape tooth will slide back along the locking face until it reaches the locking corner; at this point impulse will begin. The tooth will push its way along the impulse face until reaching the discharge corner, where it will drop until a tooth is arrested by the opposite pallet, seven and a half teeth away. Drop can be kept very small on a high-quality escapement, due to their higher concentricity,

Adjustable deadbeat pallets of an anniversary clock.

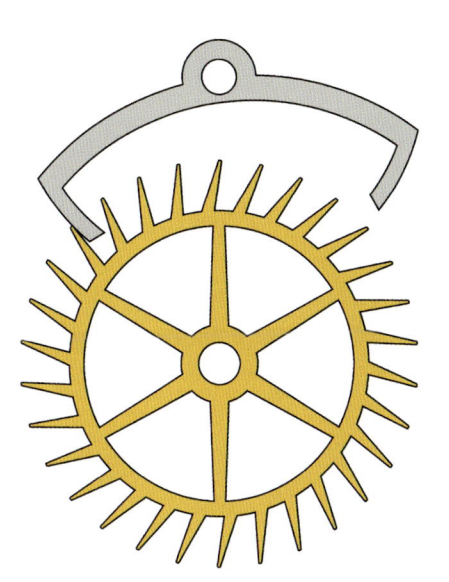

Deadbeat escapement at rest while the pendulum is at one extreme of its swing.

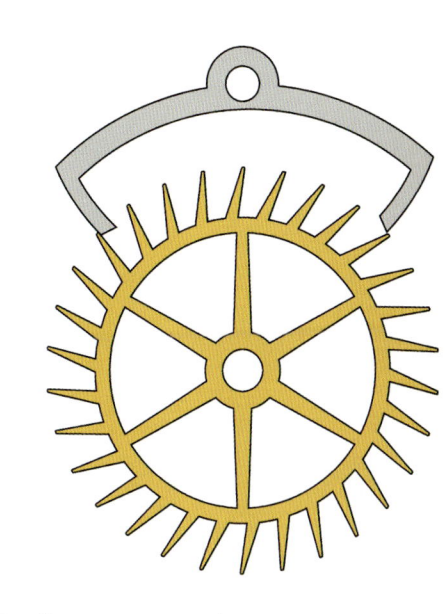

Deadbeat escapement demonstrating its lock position.

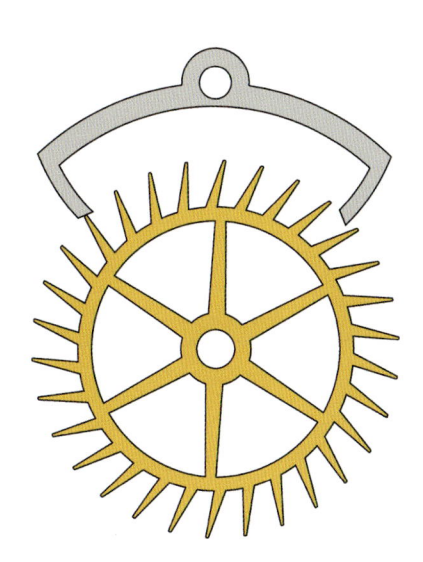

Deadbeat escapement impulsing the anchor.

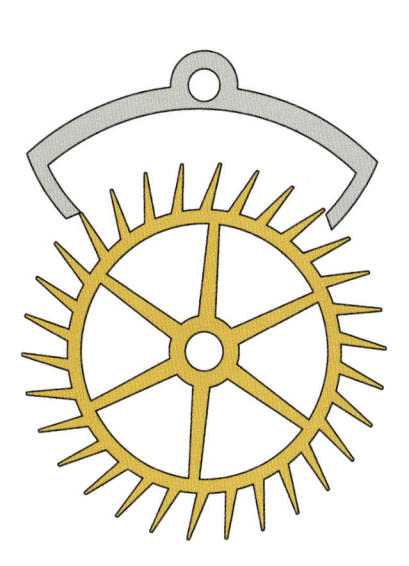

Deadbeat escapement demonstrating the amount of drop onto the exit pallet.

finer teeth, accurately ground pallets, etc., compared to a lower-quality clock, where drops will need to increase in order to allow for inaccuracies.

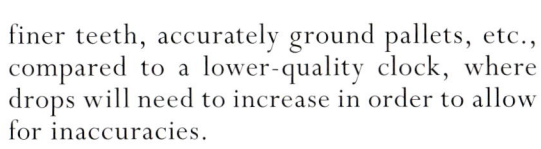

An adjustable deadbeat pallet was introduced by and named after Benjamin Louis Vulliamy (about 1820). In this configuration, curved pallets are set into the anchor and held by a screw or bracket, and are adjustable in height. By adjusting the pallets it is possible to change the geometry of the escapement by a significant amount, allowing you to grind away wear and to adjust the pallets to suit or to replace them entirely in severe cases.

56 The Timepiece

The Brocot Pin Pallet Escapement

The brocot pin pallet escapement is, when correctly adjusted, a dead beat escapement. It is often found adorning the dial centres of French clocks in its visible, decorated form, but it is similarly found between the plates, in a less decorative form.

The pallets are made of sapphire, ruby or a synthetic 'jewel', which is much harder and more wear-resistant than hardened steel but which can chip and break easily due to its brittleness. When broken, the sharp edges cut and wear the brass surface of the escape wheel. Often you will see jewels replaced with hard steel pieces which are easier to make.

The pallets are half-cylindrical and protrude forwards from the anchor, so as to intersect the escape wheel, which is mounted slightly in front. The pallets are shellacked into

Typical brocot visible escapement of a French four-glass clock.

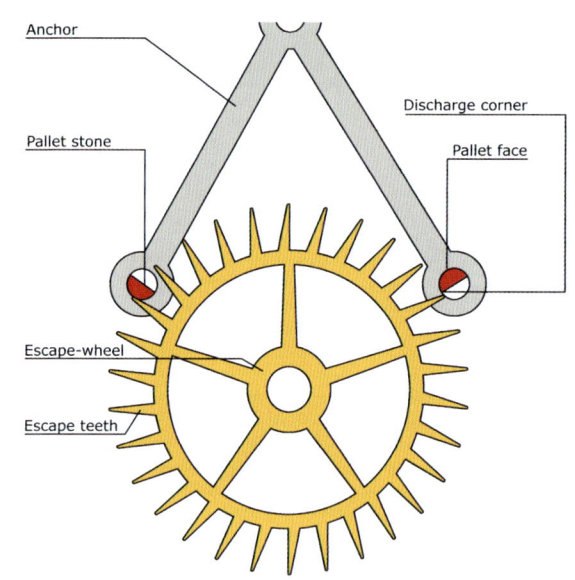

Diagram of the brocot visible escapement.

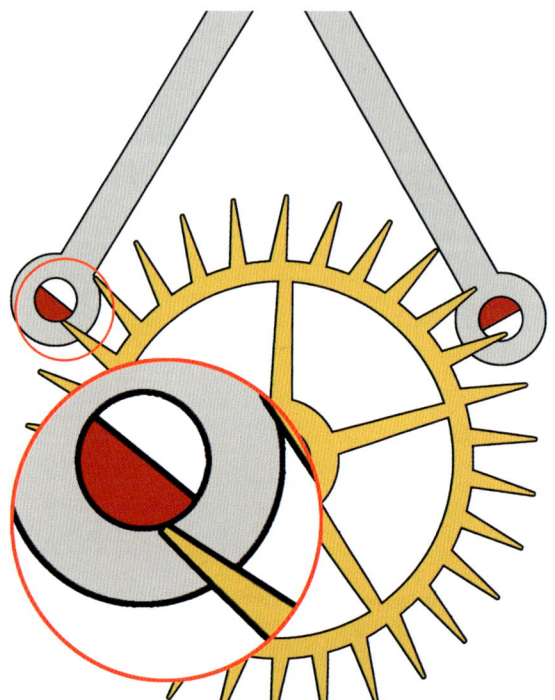

Brocot escapement demonstrating impulse on the entrance pallet.

place, or sometimes a press fit is used for those designed with steel pallets. The curve of the half-cylinder is the acting face, where locking and impulse take place. The discharge corner is where the curve meets the flat. The flat ground surface is there to provide clearance for the escape teeth, which will be made clear as you study the diagrams.

Provided the escapement is set up correctly, the escape tooth will lock at the peak of the curve, where it will rest throughout the remaining swing of the pendulum; if it locks below this curve, there will be recoil and all the related disadvantages. As the pendulum reverses, the escape wheel will unlock as the curves peak passes the escape tooth tip, which will begin to push its way along the curve of the pallet, providing impulse until the discharge corner is reached. From this point it will drop and the process will repeat on the opposite pallet.

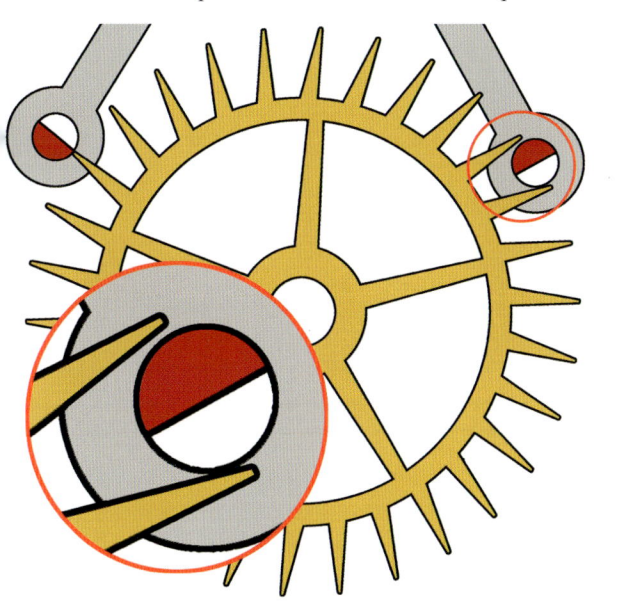

Brocot escapement demonstrating drop on the exit pallet.

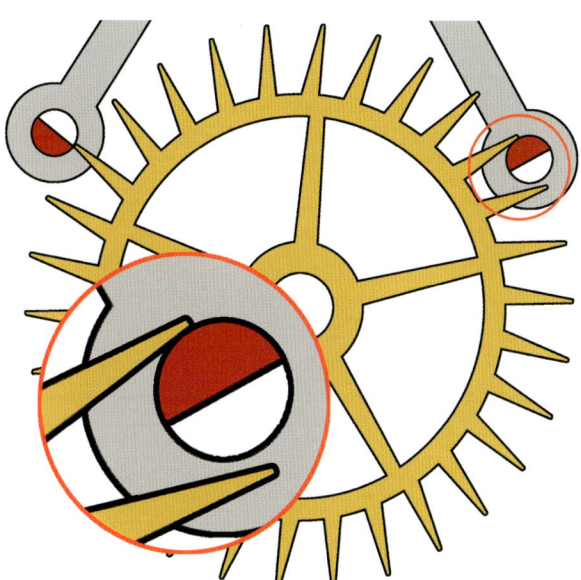

Brocot escapement demonstrating lock on the exit pallet.

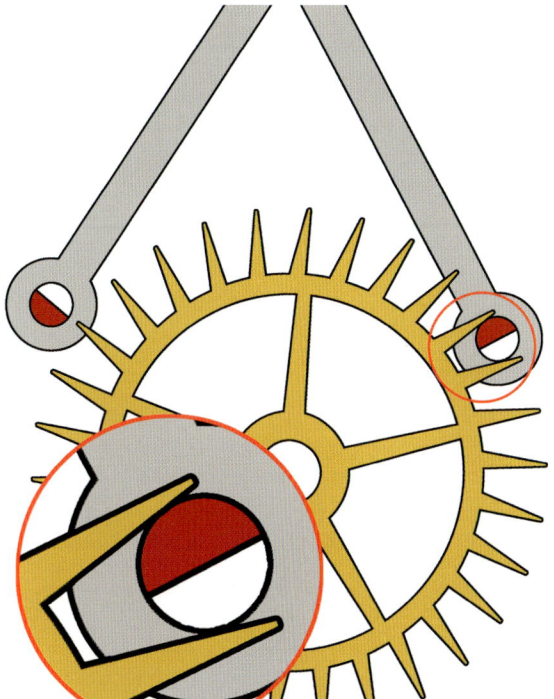

Brocot escapement demonstrating recoil/rest on the exit pallet.

Pendulum Theory

Tie a small weight to a piece of string and hold the string about 15cm (6in) from the weight; allow it to swing freely and pay attention to its speed. Now double the length of the string and let it swing again. You will see that it now swings noticeably more slowly. Next, try adding weight and repeating the experiment; you will see that there is no difference – the weight does not affect the period of swing.

The period of swing of any pendulum is a product of its theoretical length: that is, in theory, a pendulum is a mass at one end, a pivot at the other, and a weightless rod joining the two together. In reality the rod has a mass, spread over its entire length; the pendulum bob is a convex disk full of lead, its mass again spread over its full diameter. To compare the two side by side, the theoretical pendulum would be shorter than the real (complex) pendulum to attain the same rate. This is because the real pendulum has weight all along the rod, effectively cancelling out some of the weight lower down in the bob. This is why in reality a heavier bob will slow the pendulum slightly, as it 'cancels out' more of the weight of the rod and effectively lengthens the pendulum. Some pendulums have trays halfway up the rod, onto which weight can be added for fine regulation to take advantage of this effect.

Suspension

The 'pivot' of a pendulum is called the suspension, as the pendulum is suspended from it. The suspension is often a thin ribbon of spring steel, or a silk line from which the pendulum rod hangs. It allows the pendulum to swing freely without friction and can be either a separate piece or riveted to the pendulum rod. In the case of silk suspensions, the silk is tied to the back cock and around a regulating knob at the other end; it forms a U-shape for the rod to hang from, and by turning the knob you can gather up or release more silk to lengthen or shorten the pendulum.

In French clocks, a suspension type known as brocot suspension is often used. The brocot system allows for micrometer adjustment of

Silk suspension of an early French countwheel movement.

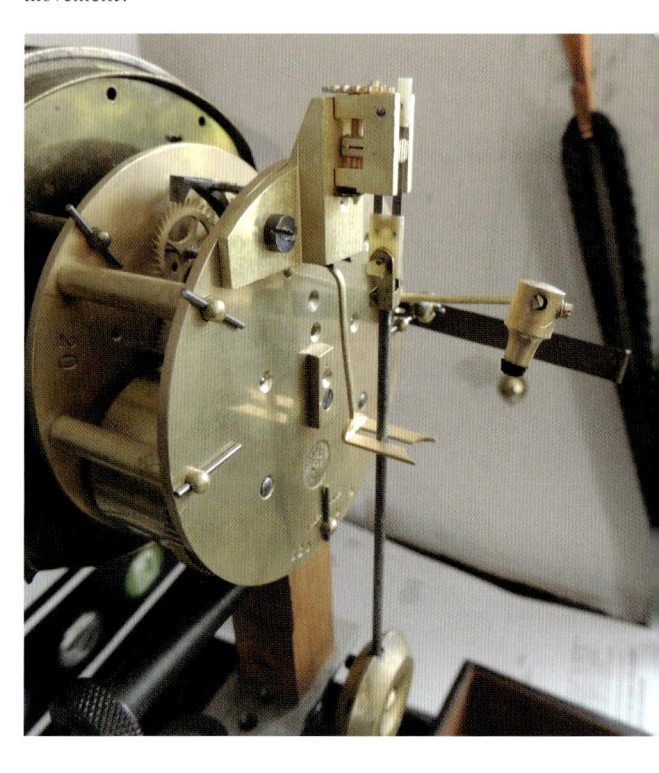

Steel suspension of a later French movement.

Brocot adjustment square of a French clock movement; this one is labelled 'R' and 'A' within the minute ring.

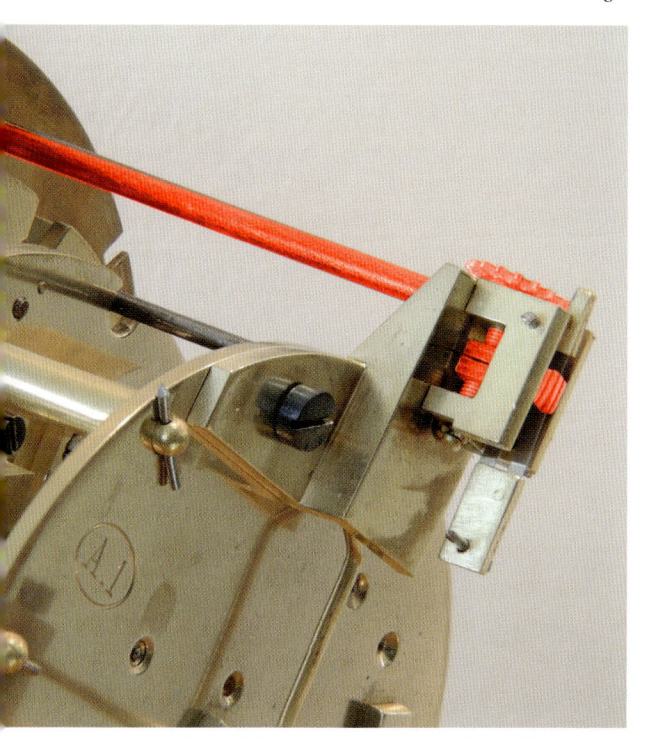

Brocot adjustable suspension unit with its functional components highlighted in red.

the pendulum via a square protruding through the dial above the XII. The suspension spring is pinned to the back cock and hangs down between two cheeks which, by being a sliding fit on the suspension, effectively alter its length by allowing only the lower half to act freely. These cheeks are mounted on a screw thread so that they slide up and down as the screw is turned through right-angle gearing by the small square that passes through the dial. By turning the square anticlockwise, the cheeks are raised, lengthening the effective pendulum length and slowing the clock.

In old English clocks, a similar arrangement is used, but the cheeks are mounted firmly; the top of the suspension rises and falls via a lever, riding on a snail or cam, attached to a small hand on the dial.

The Rod

The pendulum rod simply connects the suspension to the bob; it is threaded to allow adjustment of the bob which rests on top of the rating nut and is free to slide up and down. In more modern clocks, it is common for the rod to be made up of two or three parts which hook together; this is for ease of maintenance and does not affect the working of the pendulum. On American clocks, the rod is made of a piece of steel wire, hammered flat at one end to produce the suspension; the opposite end is bent to a hook for the pendulum bob to hang from. The rod should be as light as possible while retaining strength.

The Bob

The design of the bob is dependent on several factors. The shape is often lenticular (like a lens) and encased in a brass shell. The weight is dependent on the rod design; a heavier rod will require a heavier bob to 'cancel out' the weight high up, otherwise a longer pendulum will be necessary. Weight is also dependent on the necessary stability of the clock: in other words, a heavier bob is less easily affected by gusts of air or moving floorboards. The bob design is affected by aerodynamics but the effects are as so small as to be negligible, therefore the shape is generally down to the maker.

Typical French pendulum with its rating nut in the centre of the bob.

Typical lenticular bob of an English longcase clock, with the rating nut below the bob.

The pendulum bob rests freely on top of the rating nut. The rating nut can be below or in the middle of the bob, but the bob must rest on or hang from it so that it is able to move up and down as the rating nut is turned; the bob is lowered to slow the pendulum down and raised to speed it up.

Temperature Compensation

As temperatures rise, most materials expand. As we have seen that the pendulum length defines its rate, you can imagine that a warm pendulum will expand and slow the clock, while a cooler pendulum will contract and speed the clock up.

Many methods of counteracting this effect have been invented and a few have become commonplace. We will cover the workings of the popular varieties, and once an grasp of the principle is achieved, it will be easy to understand any type which you may come across in the future.

Mercury Compensation: Mercury bobs are used extensively in good-quality regulators as they provide a reliable compensation method which, by altering the level of mercury, is adjustable. In French clocks, the mercury is stored in sealed vials for its aesthetic qualities.

We will focus on the type of bob often found in regulators. The basic principle is that as the steel pendulum rod expands downwards with

CIRCULAR ERROR

Circular error is the effect on rate as the amplitude of swing of the pendulum increases. Pendulums swing in an arc, formed roughly on a circle because the pivot point is fixed (as close as makes no noticeable difference). A circular arc means that at higher amplitudes the pendulum takes longer to swing; it is not isochronous (when events occur at equal time intervals). In order to achieve isochronism the pendulum must swing in a cycloidal arc, the shape of which is steeper at its extremes (think more U-shaped). Attempts have been made to change the shape of the pendulum's swing by Christiaan Huygens (1670s), using special shaped cheeks for the suspension and a large amplitude. However, it has been proven that for smaller amplitudes the effects of circular error are minimal and that it is most effective to use a pendulum swing of only a few degrees, with no compensation to its arc. This is one reason why longcase clocks, with their long pendulums and recoil escapements with small arcs of vibration, were a great leap forward in timekeeping.

Mercury compensation pendulum of a French four-glass clock.

temperature increase, the mercury level rises by the same distance, keeping the theoretical length from suspension pivot to centre of mass constant. There are a few drawbacks, however: mercury is slower to react to changes in temperature than steel, so there is a small variation of pendulum length until the mercury catches up. To reduce this effect, two jars can be used, each containing half the mercury, as the smaller quantities of mercury, with their larger surface areas, react more quickly than one large jar.

The vapours from mercury are poisonous, especially for small children and especially at increased temperature; any spillages should be dealt with professionally.

Grid Iron Compensation: The grid iron pendulum is made up of nine parallel rods; four are brass and the remaining five are steel. The lengths used are representative of the thermal expansion of each material, the ratio of lengths equalling the ratio of thermal expansions. They are arranged vertically and parallel, supported by five brackets. The rods pass freely through some brackets and are rigidly attached to others, in an alternating pattern, steel rods making up the outside and centre. The steel rods always expand downwards with rises in temperature and the brass expands upwards. The expansion of brass will lift the brackets from which the steel rods hang, and the steel will lower the brackets on which the brass rods stand, and so on, cancelling out any expansion. This rod design is often faked on reproduction clocks and should not be confused with a functional rod.

Material Compensation: Wooden pendulum rods are an easy form of compensation; the thermal expansion characteristics of wood are far more stable than those of steel or brass, and provided the wood is well seasoned (so that it will not warp) and the grain is straight, the results will be very acceptable. The wood needs to be completely sealed from the atmosphere, saturated with oil or coated with varnish to keep the rod immune to changes in humidity.

Modern compensation pendulum rods are made of a nickel-steel alloy called invar, which has a thermal expansion so small that it can be ignored. In cases where finer regulation is necessary, the bob simply rests on a small brass compensating tube, which in turn rests on the rating nut. The length of the compensating tube is adjusted by trial and error until it expands upwards by the tiny amount equal to the expansion of the invar rod.

Platform Escapements

Platform escapements are controlled by a balance wheel and mounted on a plate separate from the rest of the clock movement. They are commonly used in carriage clocks and barographs, sitting horizontally on top of the movement or mounted vertically on the rear plate.

They come in many varieties ranging from cylinder escapements to chronometer escapements; however, the majority passing through the hands of the clock repairer will be either the cylinder escapement or the lever escapement.

Modern club-tooth lever escapement.

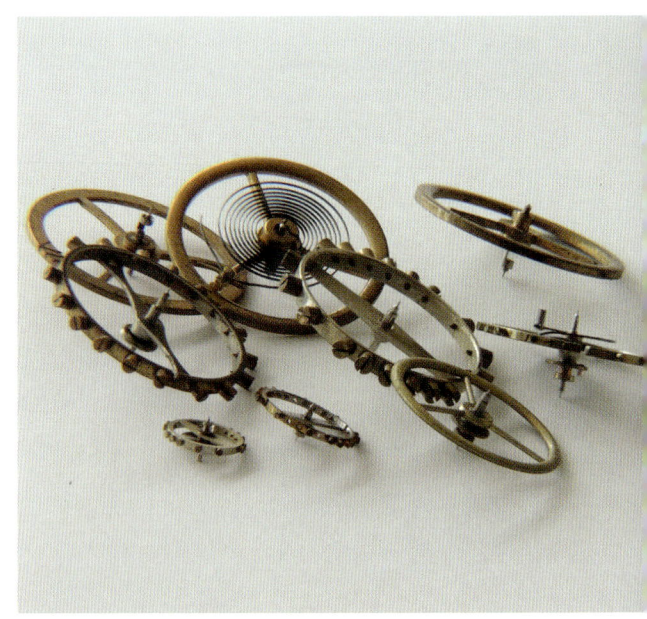

Selection of balance wheels from both cylinder and lever escapements.

Balance Wheels

Balance wheels are used to control platform escapements because of their size and mobility. They can run in any position (although timekeeping properties will vary in each position), unlike a pendulum, and can withstand movement from factors such as the swaying of a ship, which is why they are applied to marine clocks (although marine clocks are also mounted in gimbals to minimize the effects).

The balance wheel itself comes in many varieties, based partly on whether or not temperature compensation is required and to what degree.

Like the pendulum, the distance from the centre of oscillation to the centre of mass (the balance rim) does affect the timekeeping of a balance-controlled clock, and therefore so does the weight of the balance rim (as explained with pendulum theory). However, this is not the main method used to regulate such clocks; instead a hairspring is used.

Hairsprings

A hairspring or balance spring is a fine flat coil of tempered steel or a modern alloy such as nivarox (produced by a Swiss company of the same name), which has a low coefficient of expansion (and is non-magnetic and does not rust) and therefore helps reduce the effects of temperature on timekeeping. (However, alloy springs cannot be used on temperature-compensated balances, or else the compensation built into the balance will overcompensate, as this design was intended for use with uncompensated steel hairsprings.)

The introduction of hairsprings to verge pocket watches in the mid-1600s improved timekeeping significantly, and the technology continued to be developed until the financially viable introduction of the quartz crystal into horology in the 1980s. For this reason, we cannot possibly cover hairspring theory in as much depth as would be possible in a much larger publication.

The hairspring is pinned by a taper pin at its inner end to a collet which is pressed onto the balance staff so that they oscillate together, and at its outer end it is rigidly fixed, via a stud, to the balance cock. With this set-up, the balance wheel will wind and unwind the hairspring as it oscillates.

The winding of the hairspring causes the rotation of the balance to slow and eventually to stop, where the stored energy of the wound hairspring can take over and pull it back in the reverse direction. The balance wheel will gain momentum in the new direction, and receive

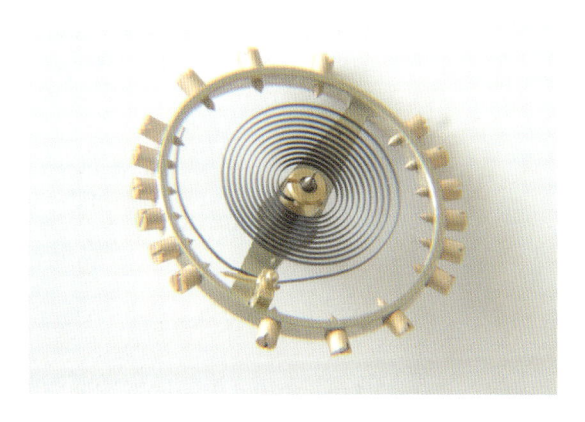

Balance and hairspring of a lever platform escapement; the hairspring collet and stud are clearly visible, as are the pins which attach the hairspring.

Lever platform escapement demonstrating the freedom of movement of the index at its extremes in this overlay image.

Fully disassembled cylinder platform escapement.

impulse from the escapement. After the spring has passed its 'at rest' position, this momentum will begin to unwind the spring, which will cause the balance to slow and reverse direction once again, and the cycle repeats.

The acting length (the length between the inner and outer attachments) of the hairspring defines the speed at which the balance will oscillate. In order to speed up the oscillation, you shorten the hairspring, and to slow it down you lengthen it. This is done at the attachment of the hairspring to the balance cock, where it is passed through a hole and pinned to the stud. To lengthen or shorten the hairspring you unpin it from the stud, extract it by the exact amount required, and re-pin it. You are then faced with an out-of-beat escapement because the 'at rest' position of the balance wheel will have changed.

To avoid all of this hassle an index lever is used. Using this system you are able to adjust the effective length of the spring without altering the resting position of the balance wheel, or unpinning the hairspring from the outer stud. It works by placing two pins against the hairspring at its outer coil. As the balance rotates, the hairspring expands and contracts, bouncing off these pins, and the effective length of the hairspring is reduced to the total distance between the index pins and the centre of the collet. The index pins can be moved along the length of the hairspring as necessary to adjust timekeeping by pushing the lever to and fro.

The Cylinder Escapement

The cylinder escapement is a frictional rest escapement and is, for that reason, inferior to the lever escapement. This is because the escape wheel rests directly on the arbor of the balance wheel (called the cylinder in this case) during rotation, producing excess friction. During the 'rest' component of the escapement cycle, the driving force of the mainspring presses the acting escape tooth against the cylinder wall, affecting the amplitude of the balance wheel, and so the escapement is also power-sensitive.

The cylinder platform escapement consists of the following items:

- The main plate or platform onto which the other components are attached; this contains

only clearance and screw holes cut into its surface.
- The carriage, which fits flush into the main plate and contains the lower jewel-hole for the balance; it also contains mounting points for the balance cock and lower balance end stone.
- The balance cock provides the upper jewel-hole for the balance and carries the upper end stone, the index and the hairspring stud hole.
- The index pivots concentrically with the hairspring and is a friction fit between the balance cock and the upper end stone; it provides two index pins (or a pin and 'boot'), which act upon the hairspring, and a lever for adjustment.
- The balance assembly contains the cylinder itself, an uncompensated balance wheel and the hairspring.
- Finally, you have the escape wheel and the escape wheel cocks, one of which is slung below the main plate and carries the lower jewel-hole and the other secures the wheel from above and carries the upper jewel-hole. A lower end stone is sometimes used here to reduce friction.

Step by step, the process is as follows. A tooth of the escape wheel rests on the outer wall of the cylinder as the balance rotates. When the opening of the cylinder approaches the tooth tip, the escapement unlocks and the sloped face of the escape tooth begins to push the cylinder's edge out of its way, providing impulse. When the end of the tooth is reached, the wheel is free to drop onto the inside wall of the cylinder, while the balance continues to rotate. The shape of the cylinder cut-outs allows a full amplitude to be achieved without the cylinder and escape wheel colliding. The cylinder escapement does not recoil, and therefore goes through the same drop, lock, rest and impulse sequence as any other deadbeat escapement, the main difference being that impulse takes place mostly on the surface of the teeth, rather than on the impulse face of the 'pallets' as in the deadbeat anchor escapement.

The escape wheel is an unusual design in that the acting surfaces of the teeth are raised above the flat of the wheel. This is to provide clearance of the cylinder as the balance rotates. The cylinder itself is cut with a clearance slot to allow the escape wheel to pass, while allowing the raised tooth to act on its inside surface.

The cylinder is a hard steel tube (sometimes jewelled cylinders are used but we will not go into that) into which cut-outs are ground; steel plugs are pressed into the ends of the cylinder for the pivots to be turned onto; the cylinder is then pressed into the centre of the balance wheel; and the hairspring collet is fitted above the balance wheel.

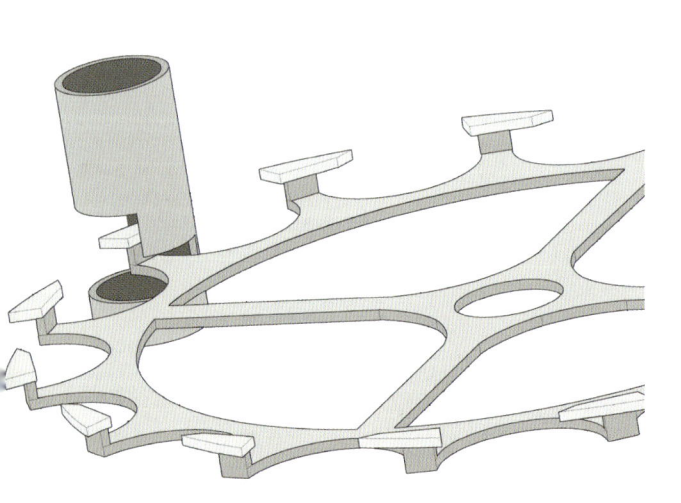

Clearance slot cut into the cylinder wall to allow the safe passage of the escape wheel as a tooth is at rest.

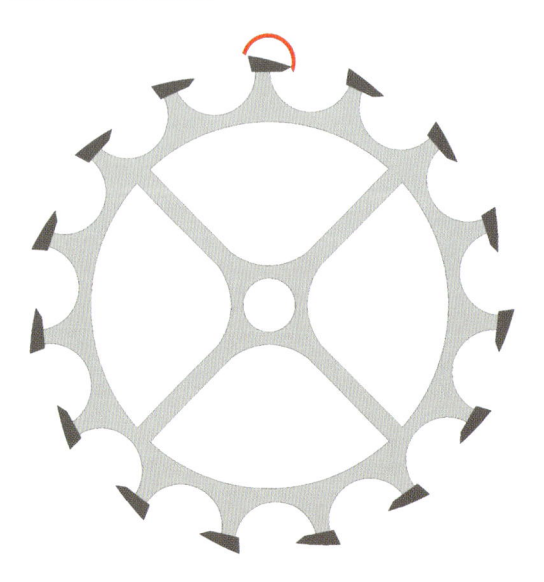

Cylinder (red) with an escape tooth locked on its inside surface.

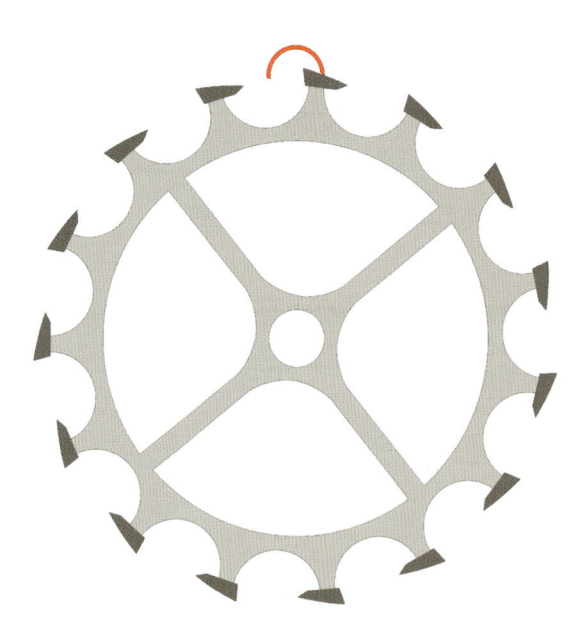

Escape tooth exiting the cylinder (red) and providing impulse.

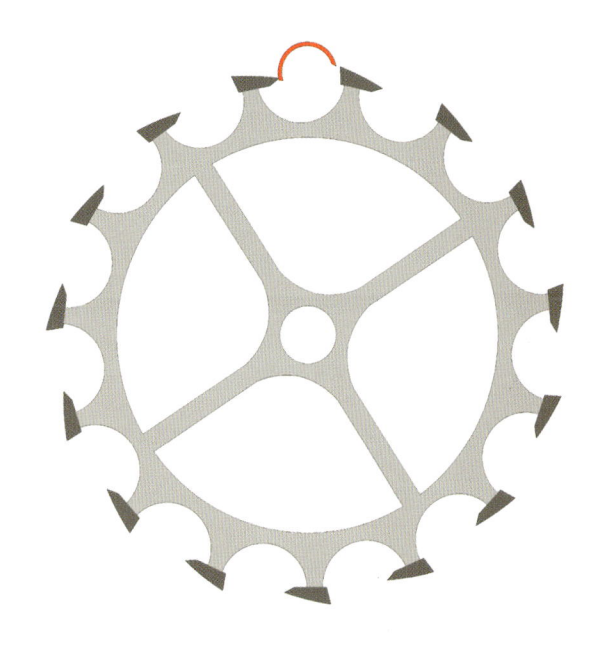

Escape tooth locking onto the outer surface of the cylinder wall (red).

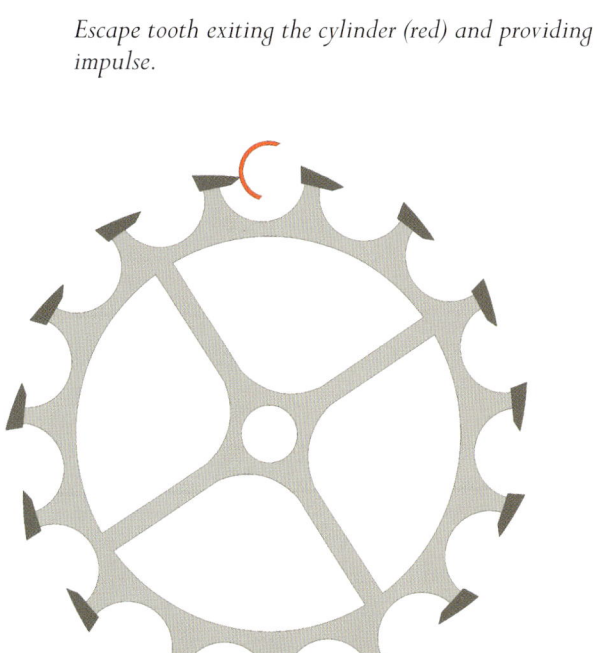

Cylinder (red) in frictional contact with the escape wheel throughout the majority of its oscillation.

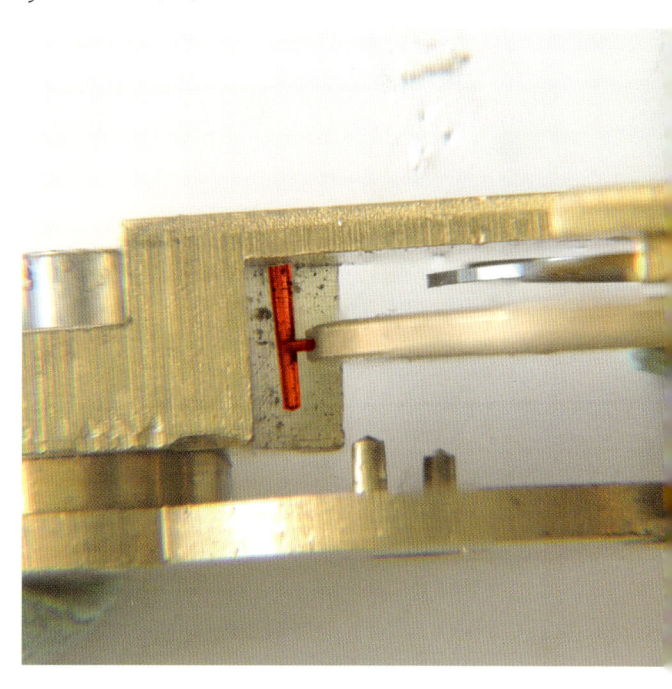

Safety pins of the cylinder escapement (highlighted in red); they should never come into contact under normal operation.

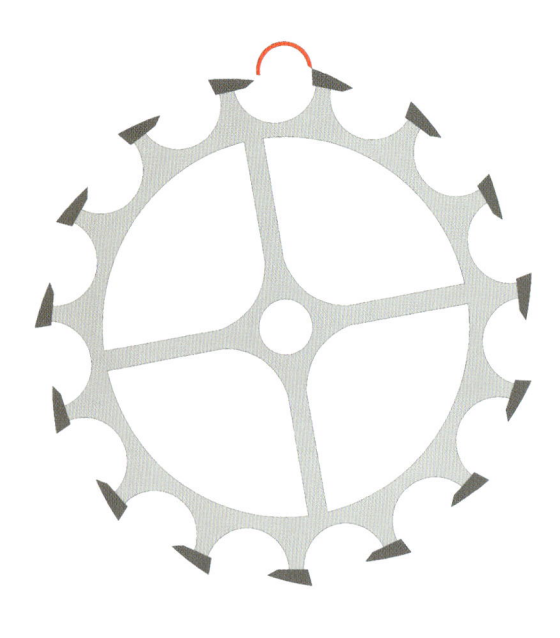

Demonstration of a cylinder which is set shallow to the escape wheel; note how the escape tooth will always land on the entrance to the cylinder and never lock safely.

The carriage and balance cock are a separate assembly, allowing changes in the depth of the escape wheel to cylinder.

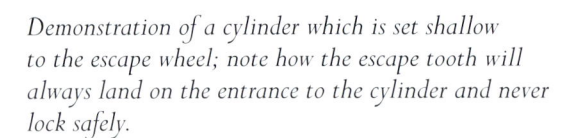

At rest, the balance wheel has a pin projecting forwards from its rim, and the balance cock has a pin projecting downwards 180 degrees away, to intersect it and provide a safety action for the escapement. As the balance wheel rotates, these two pins will collide, (at 90 degrees to each other) protecting the cylinder from damage.

If this action were to fail, the cylinder clearance slot would crash into the escape wheel if excessive balance vibration were to occur.

The cylinder escapement can be adjusted for depth of engagement between the escape wheel and the cylinder. The carriage carries the full balance assembly and is free to move within the main plate, which carries the escape wheel. By loosening the carriage screw you are able to move the cylinder deeper or shallower in engagement. If a cylinder is too shallow in engagement to the escape wheel, then the impulse face of the teeth will land on the entrance of the cylinder, causing it to run through at great speed.

The carriage is pinned in place during manufacture to avoid slippage from its original position, so it is necessary to reduce the diameter of these pins by filing beforehand. This should only be necessary once the escapement has begun to wear, so it is an alternative to replacing parts which in many cases would cost in excess of the value of the clock.

The Lever Platform Escapement

There are several variations of lever escapement in use, but the two most common in clocks are the English straight-toothed lever escapement and the Swiss club-toothed lever escapement. They are similar enough in design that at this time you only need to differentiate between the impulse action, which for the English lever acts purely on the pallet stones and for the Swiss lever is split between the pallet stone and the clubbed tooth of the escape wheel. We will focus on the Swiss club-toothed lever escapement as it is the modern standard.

PIVOT SHAPE

There are two pivot shapes commonly used in escapements; they both serve different purposes and work in different ways, and the type of pivot used defines the type of pivot hole which must be used.

In balance staffs and cylinder plugs, a tapered pivot shape is used; these require the use of end stones and greatly reduce friction as well as increasing the strength in the pivots. The pivot has a much shorter parallel portion at its end than a standard straight pivot. A thin parallel pivot is a weak point in any arbor and a common point of failure; by keeping this small and gradually tapering out towards the wider part of the arbor, you greatly increase the strength of the pivot.

With a tapered pivot you need to ensure that the pivot cannot pass freely through the pivot hole, or else as the pivot widens, it will jam in the hole and crack or chip the jewel. To ensure only the parallel portion enters the jewel-hole you use an end stone; this is a second jewel which caps the hole.

The jewel-hole itself is an unusual design when an end stone is used; it has the appearance of being upside down, with the oil sink facing towards the arbor rather than away. This design allows for best retention of oil when an end stone is in place.

Straight pivots are often (though not always) used for escape wheels and pallet arbors in platform escapements. In this design there is a clear shoulder between the pivot, which has parallel sides for its entire length, and the arbor. The shoulder must be perpendicular to the pivot and with a sharp corner where they meet. During operation, the shoulder rests against the jewel (bottom jewel only due to gravity); if there is any obstruction or widening of the pivot where it meets the shoulder it will jam in the jewel-hole and cause damage.

The jewel-hole in this design has the oil sink on the outside, and a flat polished surface for the shoulder to rest on the inside.

Sometimes an end stone is used on the lower pivot of the escape arbor or pallet arbor even though a straight pivot is used. This is to reduce friction caused by the shoulder resting on the jewel; no upper end stone is necessary.

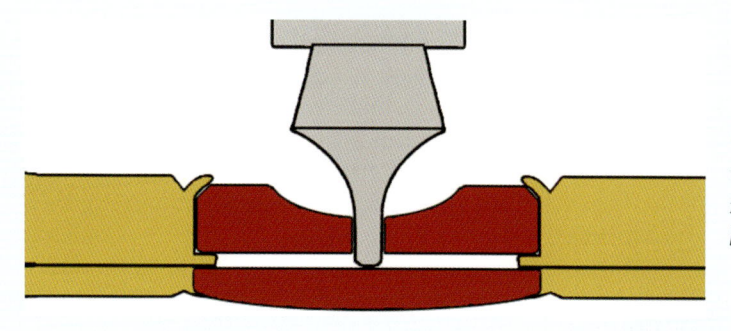

This pivot shape is used only when paired with an end stone; note how the oil sink appears to be upside down.

Standard pivot shape for most arbors; note the rubbed-in jewel setting in this cross-section.

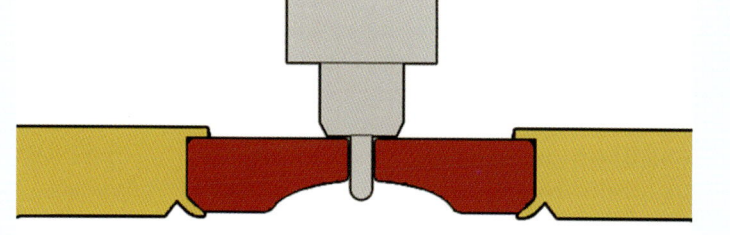

Fully disassembled lever platform escapement.

Swiss club-tooth lever platform escapement.

The lever escapement is a detached escapement, meaning that for the majority of its rotation, the balance wheel is in no way mechanically connected to the escapement. This allows the balance to rotate uninterrupted for most of its vibration, with the exception of the effects of the hairspring.

The balance's arbor is called the balance staff. As well as carrying the balance and hairspring, the balance staff also carries the roller.

Either the roller and safety roller, (the double roller), or a single roller can be used for all these functions. Their job is to provide a 'passing hollow' for the guard pin of the lever and to hold the impulse jewel.

Impulse jewels come in many shapes and sizes but modern clocks and watches have settled on a D-shaped jewel as in the diagrams here. These D-shaped jewels are cylindrical with one side ground flat. This allows the impulse jewel to enter the 'horns' of the lever with the most efficient clearance to provide strong and even impulse. The impulse jewel is held in the roller with shellac, and must be perfectly upright and of exactly the right size: too large in diameter and it will not enter the lever notch; too small and there will be excessive freedom and lack of impulse. If it is too short it will (again) not enter the lever; if too long it will foul the guard pin.

The lever has two ends; at one end are the pallets and at the other are the horns and guard pin. At the horn end of the lever a safety action is necessary to stop the escapement from unlocking when the clock is shaken or knocked: this would stop the clock and possibly damage the escapement. The safety action comes in three parts: the guard pin and safety roller, the design of the horns, and draw.

Draw is a function unique to the lever escapement, ensuring that the lever is pulled away from the balance staff when at rest to reduce friction. It works by setting the locking face of the pallet at an angle to the escape tooth. Imagine a force representing the direction of rotation of the escape wheel, and a wall constrained to move only up or down representing the pallet stone. When the force is at 90 degrees to the wall, there will be no movement of either part. If the wall leans in to the left, then the force will tend to push the wall away and up as it slides along its surface, or to impulse. If the wall now leans to the right, the force will tend to climb up the wall or to pull the wall downwards, or draw it in.

The guard pin is part of, or attached to, the lever. Its length is important to allow adequate clearance of the safety roller while maintaining a safe action. While the passing hollow (a cut-out which allows the guard pin to pass the roller during impulse) is out of reach, the guard pin stops the pallets from unlocking and impulsing the lever across to the opposite banking pin before

Double roller with its passing hollow on a different level from the impulse jewel; note also the D-shape of the jewel.

Single roller with its passing hollow ground on the same level as the impulse jewel; note the round impulse jewel of this older balance.

Horns and guard pin of the lever escapement.

70 The Timepiece

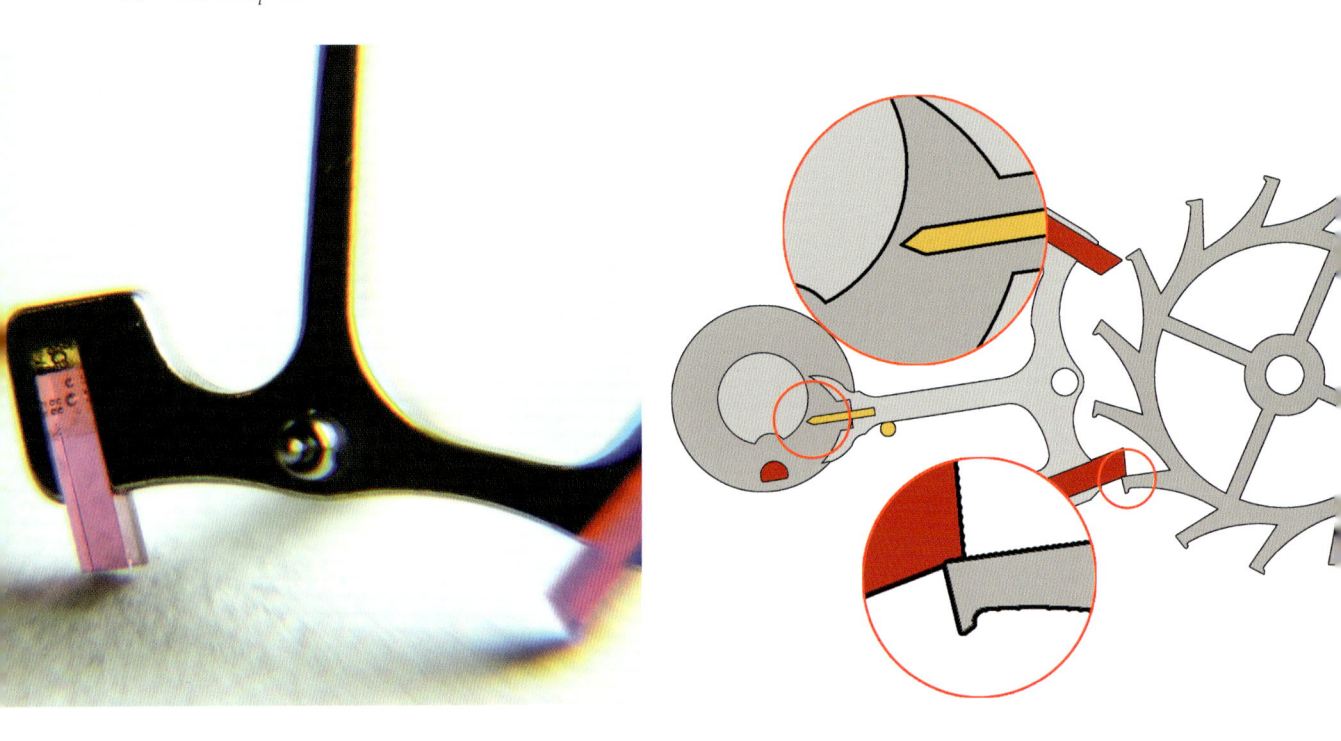

Pallet stones of the lever escapement; note the shellac used to hold them in place.

Safety action of the lever escapement; note how the guard pin will contact the roller before the escape tooth unlocks.

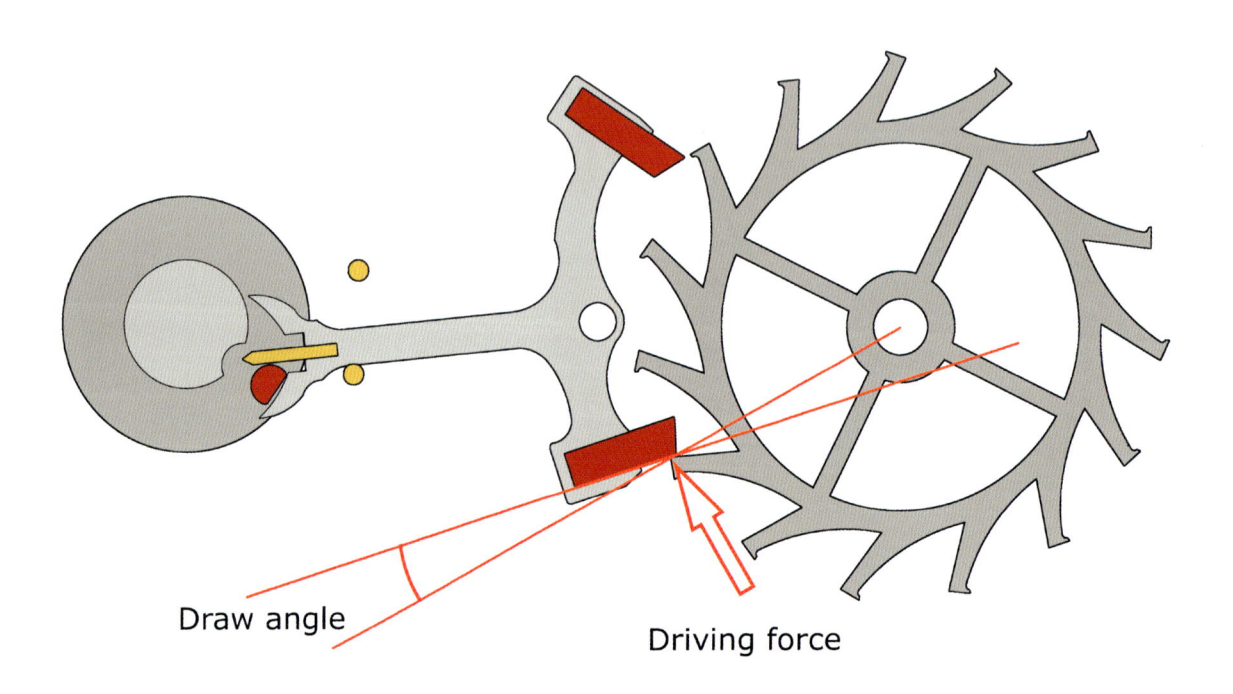

Draw angle

Driving force

Draw angle of the lever escapement works with the safety action to keep the balance oscillating freely.

the balance is in position to do this itself, consequently causing 'overbanking' (when the impulse jewel crashes into the outside of the horns).

When the clock is knocked, the guard pin hits the edge of the safety roller. This movement must be less than the total depth of lock and run to banking (the distance through which the pallets are drawn between locking and contact of the banking pins), or else the lever will no longer be drawn against the banking pins but will unlock and be pushed hard against the safety roller as the escape wheel tries to impulse the pallets.

Meanwhile, the horns are shaped to allow the impulse pin to pass freely, while at the same time creating a safety function for the moment when the guard pin and the passing hollow first coincide. When a knock is received in this position, the guard pin will enter the passing hollow, and the horn will contact the impulse pin and then be drawn away. This is not ideal, but it is for such a brief period of time that it is acceptable.

The functioning of the escapement is fairly complicated to explain as you have to pay attention to both ends of the lever to fully understand what is going on. We will first cover the horn end of the lever through one vibration of the balance, and then we will focus on the pallets and escape wheel.

As the balance is at one extreme of its vibration, and the lever is drawn in hard against the banking pin with the escape tooth locked, the direction of balance rotation reverses and the impulse pin begins to advance towards the notch of the lever. The impulse pin approaches the lever and passes the curve of the horn with freedom before it enters the notch. As it starts to move the lever, the guard pin enters the passing hollow. The lever continues to rotate with the balance until the pallet unlocks, when the lever flicks forward and the freedom of the impulse pin in the notch allows the lever to 'drop' onto the impulse pin; at this point the balance no longer moves the lever, but the lever impulses the balance. The lever, which is now being advanced by impulse from the escape wheel, pushes against the impulse pin and passes that impulse to the balance wheel. When the lever locks and is drawn against the banking pin, the impulse jewel exits the notch, passes

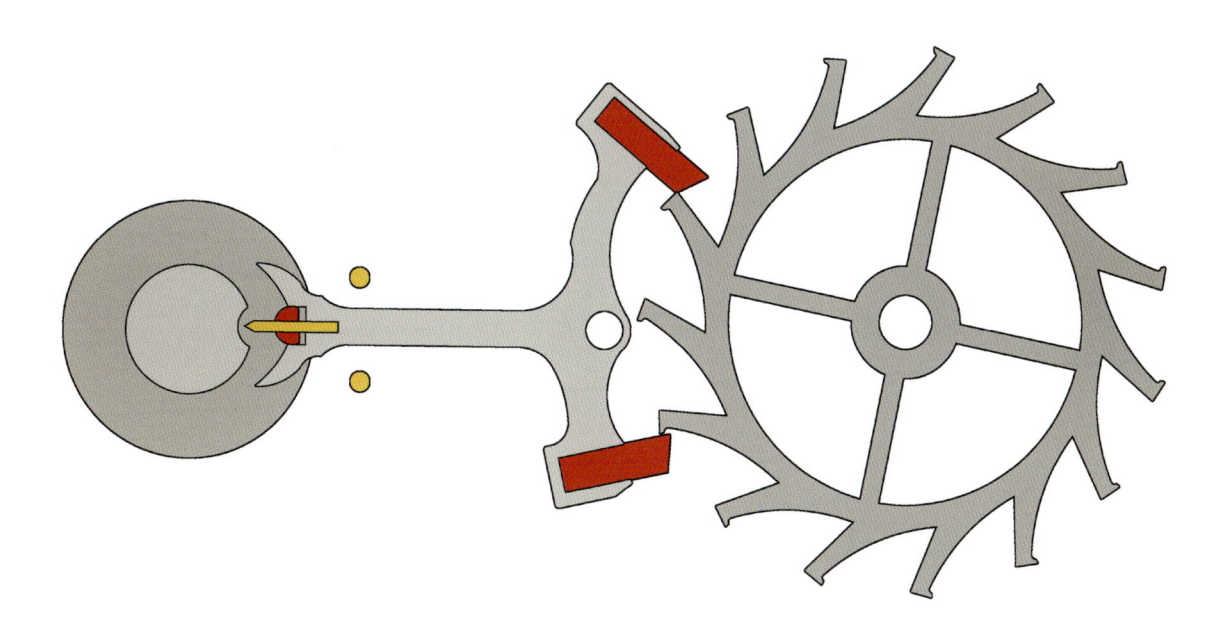

Lever escapement about to unlock on the exit pallet.

72   The Timepiece

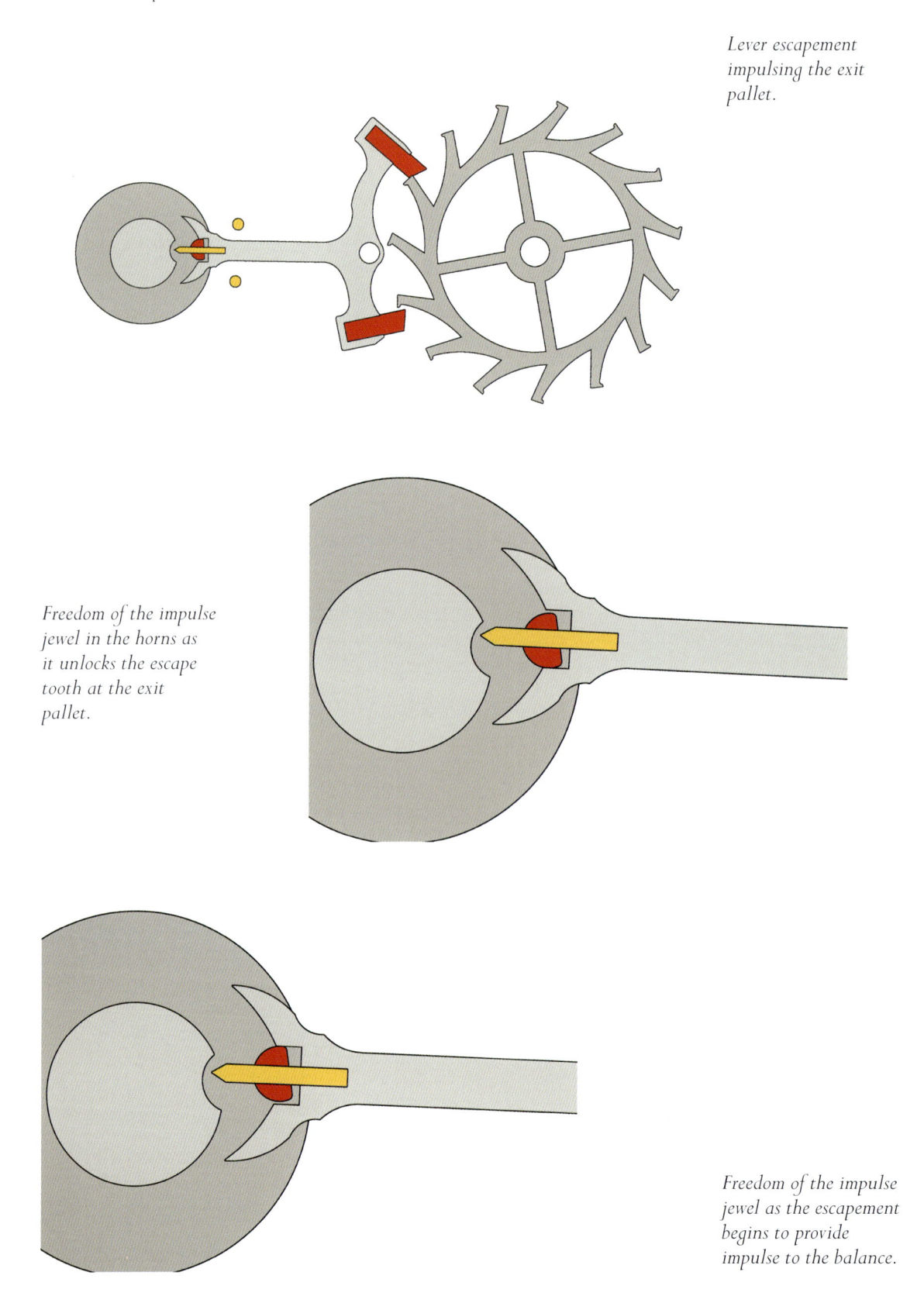

Lever escapement impulsing the exit pallet.

Freedom of the impulse jewel in the horns as it unlocks the escape tooth at the exit pallet.

Freedom of the impulse jewel as the escapement begins to provide impulse to the balance.

An escape tooth locks onto the entrance pallet.

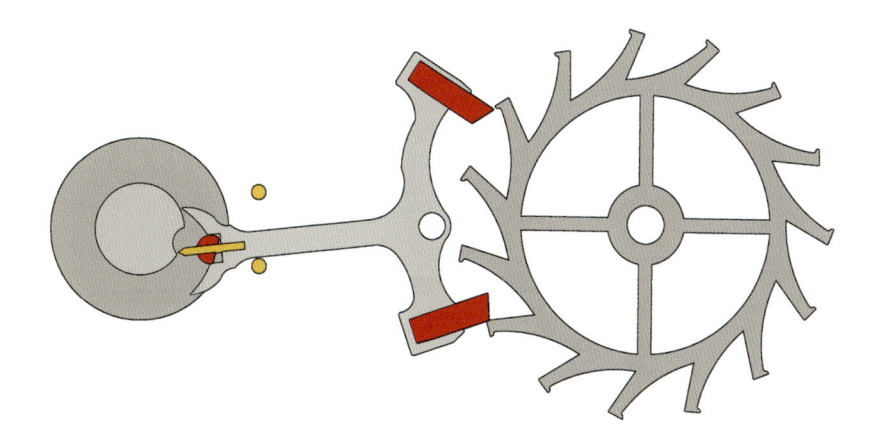

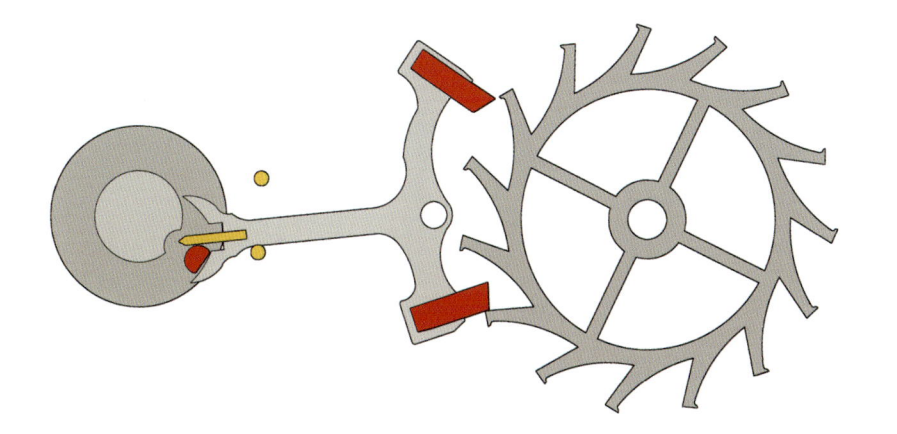

Lever escapement fully locked and drawn against the banking pin.

freely by the opposite horn and the balance continues its vibration unimpeded.

At the pallet end of the lever you see the following process. As the impulse jewel begins to move the lever, which is being drawn against the banking pin by the escape wheel, it causes the pallet to lift and the escape wheel to turn backwards by a tiny amount as the draw angle is overcome; this continues until the locking corner of the pallet stone is reached by the tooth and the escapement unlocks. Now the escape wheel rotates and pushes the pallet out of its path impulsing the balance. As impulse ends, and the impulse pin exits the notch, the escape wheel drops safely onto the locking face of the opposing pallet stone, and the force of the escape wheel rotation draws the pallet stone downwards and towards the escape wheel centre. The amount of draw defines the 'run to banking', which is set by the position of the banking pins. The amount of run to banking must be even for each pallet otherwise the action of the escapement will be inconsistent, as unlocking one pallet will require more power caused by deeper draw, or the safety action will fail due to shallower draw.

The balance amplitude of the Swiss lever escapement is 275 to 315 degrees, best observed by watching the impulse pin.

Chapter 5

Striking Clocks

A striking clock sounds the hour, and sometimes the half-hour, on a gong or bell. In its most basic form, striking is achieved by lifting a hammer and dropping it on the bell or gong repeatedly until the number of hours has been struck. There are two different ways to achieve this, rack and snail striking and countwheel striking.

In order to strike reliably a strike train must achieve three things:

- prepare the strike train and provide adequate clearances prior to striking
- provide a way of counting the number of blows to be struck
- repeatedly lift and release the hammer at evenly spaced intervals.

BELLS AND GONGS

Clock bells are small cast domes of bell metal, a type of bronze which produces a consistent note when struck, provided that they are free to vibrate. For this reason the bell is suspended from its centre by a bolt passed through a hole, allowing its circumference to vibrate without obstruction. When a bell casting cracks, which is not uncommon, then the two faces of the crack vibrate against each other and obstruct

Typical bell-striking French clock.

Good-quality English bracket clock with gong strike.

VARIETIES OF STRIKING CLOCK

Dutch Striking

A less common form known as Dutch striking uses a very similar system to strike the hour on a bell in the same manner as standard twelve-hour striking (one blow for one o'clock, two blows for two o'clock, and so on). It differs in that at the half-hour, it will strike on a second, higher-pitched bell for the full number of the upcoming hour. Although it is called Dutch striking it is also found in clocks of other origins.

Roman Striking

The aim of Roman striking is to reduce the number of blows needed to sound the hour and therefore allow for increased duration of the clock. For this reason it is often found in month-running longcase clocks (to achieve the same duration from a standard striking mechanism, a much larger weight would be needed to cope with the increased gear ratio).

Roman striking works by assigning one low-toned bell the numeral value of V or five, and the other, higher-toned bell the numeral value of I or one. The clock then proceeds to strike in Roman numerals. A single blow on the deeper bell indicates five o'clock rather than five separate blows; two strikes of the deeper bell (V + V = X) followed by two blows of the higher-toned bell indicate twelve o'clock.

Although clock dials tend to show IIII at four o'clock, Roman striking will strike IV rather than IIII.

Ship's Striking

Ships' clocks often strike one blow for each half-hour that has passed, until eight strikes have been reached; for example, at midday the clock will strike eight blows rather than twelve, and one blow at twelve thirty, but it will strike two blows at one o'clock, three at one thirty, and so on until eight blows is reached once again at four o'clock, where the process will reset. This divides the day into four-hour shifts.

Trumpeters

A trumpeter clock is a type of clock made in the Black Forest region of Germany, and appears similar to a cuckoo clock in design. Bellows are used to blow air through a small 'trumpet' at the hour. Complex pinned musical barrels are able to play multiple trumpets to produce incredibly complex tunes.

---

the free vibration of the bell as a whole, resulting in a dull thud.

Gongs are used as a cheap alternative to bells or to produce a much deeper tone. They are often a coil of blue steel suspended from one end or a straight wire of bronze as in most quarter-chiming clocks. The end must be firmly mounted to a soundboard (as the strings of a piano are) in order for the vibrations to be heard. In most cases the soundboard is simply the case of the clock.

The mechanical operation of the bell and gong striking are very similar, although geared so that the striking of a gong is slower. From this point on I will refer to both mechanisms as striking a bell for simplicity.

COUNTWHEEL V. RACK

There are two types of strike mechanism, countwheel striking and rack and snail striking. The following is a brief description of their differences.

Countwheel striking was the earliest type of striking mechanism. It requires a countwheel

Rear view of a countwheel-striking French movement; note the detent resting within the countwheel.

Early countwheel, with twelve labels for the hours struck.

mounted so that it turns with one of the slower wheels of the gear train (often mounted to the rear of the second wheel of French clocks or the greatwheel of early longcase clocks).

The countwheel itself has various cut-outs around its circumference, each of which is a stopping point to lock the strike train via a detent, while the raised pieces between them represent the hour to be struck. You will notice that the space between cut-outs increases in length respective to the previous gap; the amount it increases represents the time that the countwheel is allowed to rotate for a single strike of the bell, therefore adding a single blow to each section.

The countwheel is sequential: it will always strike in increasing order, one, then two, then three and so on. The problem here is that a single accidental or missed strike will have the clock striking the wrong hour from that point onwards (if the clock fails to release the strike at four o'clock, it will then strike four at five o'clock, five at six o'clock, etc.). After a few years as the movement begins to collect dust and its oil evaporates, strike failures may become more frequent and the clock will forever need manually resetting or regular overhauls.

There are many very nice, and valuable, countwheel clocks; it is not that they are components of a lesser mechanism simply because the design was an early one; however, the use of a countwheel alone is not an indication of age or quality. The French continued to use them frequently well into the existence of its successor, the rack and snail. They are also commonly found in mass-produced clocks of the twentieth century.

Because of the missed strike problem countwheel striking was superseded by rack and snail striking. In this form of striking mechanism, the component which counts the number of blows to be struck is directly attached to the hour hand. Therefore when properly set up, a rack-striking clock will always strike the same number of blows as the hour shown on the dial, even when a single strike has failed or been accidentally released.

The invention of rack striking is attributed to Edward Barlow, and was first put to use by Thomas Tompion in the 1670s. The number of blows is controlled by the combination of the rack and snail. The rack can be considered as a segment of a large wheel (though the tooth form is not of that used in gearing) and spans twelve teeth, one for each hour. Prior to the hour being struck, the clock 'warns' and the rack is released and is free

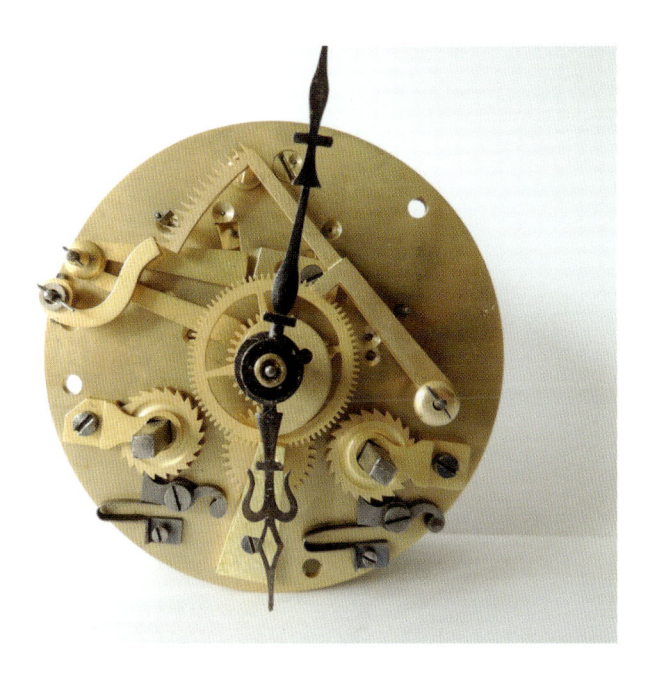

Front plate of a French rack-striking clock.

Front plate of a repeating carriage clock demonstrating a flirt release snail.

to fall; the amount by which it is allowed to fall is controlled by the snail, which is a snail-shaped cam wheel. The rack will fall to expose twelve teeth at twelve o'clock, one at one o'clock and so on. One tooth will be gathered up for each blow of the hammer.

On repeating clock movements it is important for the clock to strike the current hour when requested to strike. Using the standard snail design this is not possible, so a flirt release snail is used. This removes the snail from the hour wheel and mounts it separately on a post. It is held in place by a star wheel and finger arrangement, which will hold the snail in position and allow it to jump to the next division when it is pushed around by a pin. The pin is mounted on the cannon pinion or minute wheel, therefore flirting the snail around to the next hour just before the clock warns.

WARNING

A few minutes before the hour, the clock 'warns', which is a phase where all striking mechanisms prepare themselves. There is a wheel in the gear train dedicated to this function: the warning wheel is second to the fly at the top of the strike train, it has a single pin protruding from its surface which is to be held up after warning has taken place, and it is by releasing this pin that the clock is allowed to strike. The warning of the strike train is important because it allows all locking detents, such as that of the hoop wheel (on a countwheel-striking clock) or the gathering pallet's tail (on rack-striking clocks) to be completely free and unable to re-lock when the strike is released. It also ensures that the rack has had time to fully drop onto the snail before striking commences.

It is important during warning that roughly half a turn of the warning wheel is achieved (to observe this, the warning pin should be 180 degrees from the detent of the lifter when the strike is locked), both to guarantee safety of the unlocking action and to ensure that the bell hammer does not begin to lift prematurely.

There are occasions where warning is unnecessary. For example, repeating carriage clocks are designed to ensure safe unlocking at an instant and, when manually operated, the fly itself is held up during the button press in order to allow the rack to drop; this is sufficient provided a brief pause is provided by the operator.

Warning mechanism of a French movement in the locked position (pin and detent highlighted in red).

Warning mechanism of a French movement in the warned position (pin and detent highlighted in red).

COUNTWHEEL TRAIN

There are two main variations of countwheel gear train, one commonly used by the English and one used by the French. The differences are minor, and as French movements are more common, we will focus mainly on these, with an aside for the English variety.

Any gear train starts with a power source, which can be any of the sources previously described; however, in the average French countwheel movement, the power source is a going barrel.

The greatwheel drives the second wheel pinion, whose arbor extends through the rear plate and which carries the countwheel. The second wheel is riveted to its pinion and sits behind the barrel; it meshes with the pinwheel arbor pinion. On English countwheel clocks, the rear pivot of the barrel arbor is often extended through the back plate to receive a small gear to drive the countwheel; alternatively, the countwheel is riveted directly to the side of the greatwheel.

The pinwheel is a wheel with a series of equally spaced steel pins protruding forward from its

Front plate of a French rack-striking clock in the warned position; note how the rack has fallen and been caught by the snail.

rim; these pins lift and release the hammer as the wheel rotates. Its rear pivot is in a cock screwed to the back plate, rather than in the back plate itself. It is important that the pinwheel is in the correct register with the next pinion up the train; by removing the cock on the back plate you are able to rotate the wheel to the correct position without parting the plates. The pinwheel meshes with the pinion of the locking wheel arbor.

The locking wheel has a single pin protruding from its rim: it is this pin that locks the strike mechanism when the countwheel detent drops into one of the cut-outs, taking with it the locking detent. The locking wheel meshes with the warning wheel arbor.

The locking wheel of English countwheel clocks is called the hoop wheel. Instead of a locking pin the wheel has an incomplete hoop of brass riveted to its rim, the gap allowing the hoop-detent to drop and lock on the leading edge of the hoop itself. Unlike the French variety the countwheel-detent does not ride along the countwheel but periodically 'feels' for a cut-out. As the gap in the hoop is reached the countwheel-detent drops and 'feels' the countwheel; if a cut-out is felt then the hoop-detent will catch the leading edge of the hoop and lock the strike, otherwise the countwheel will not allow the hoop-detent to pass into the gap of the hoop and so the strike cannot lock.

The warning wheel, again with a pin protruding from its rim, is next in the sequence, and allows for the safe unlocking of the strike train. It is important that it is in the correct register with the locking wheel arbor. The warning wheel itself meshes with the fly arbor.

The fly arbor has no wheel; instead the arbor has a small groove turned into it which registers the friction spring of the fly. The fly is an airbrake which is a friction fit to the arbor; this slows the strike train to an acceptable speed, and the friction drive reduces the potentially damaging forces of locking and unlocking the strike train forty-eight times a day.

RACK-STRIKING TRAIN

The rack-striking train does not vary significantly from clock to clock. The rack and snail are mounted on the front plate of the movement, and the train provides the pinwheel to raise and drop the hammer, the gathering pallet

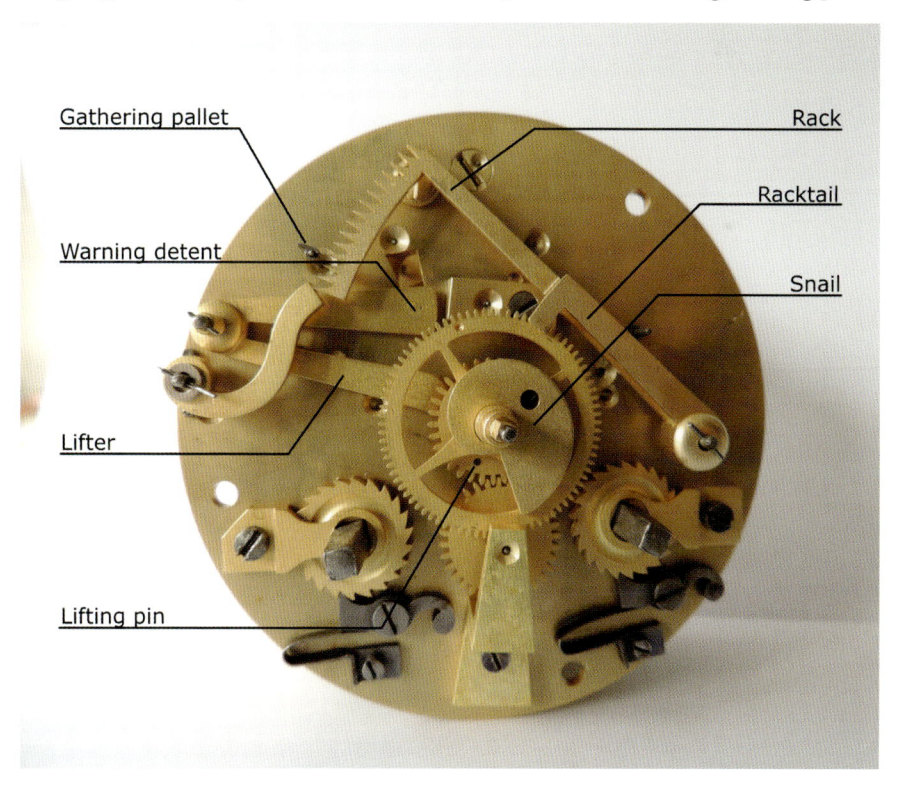

Labelled front plate of a French rack-striking movement.

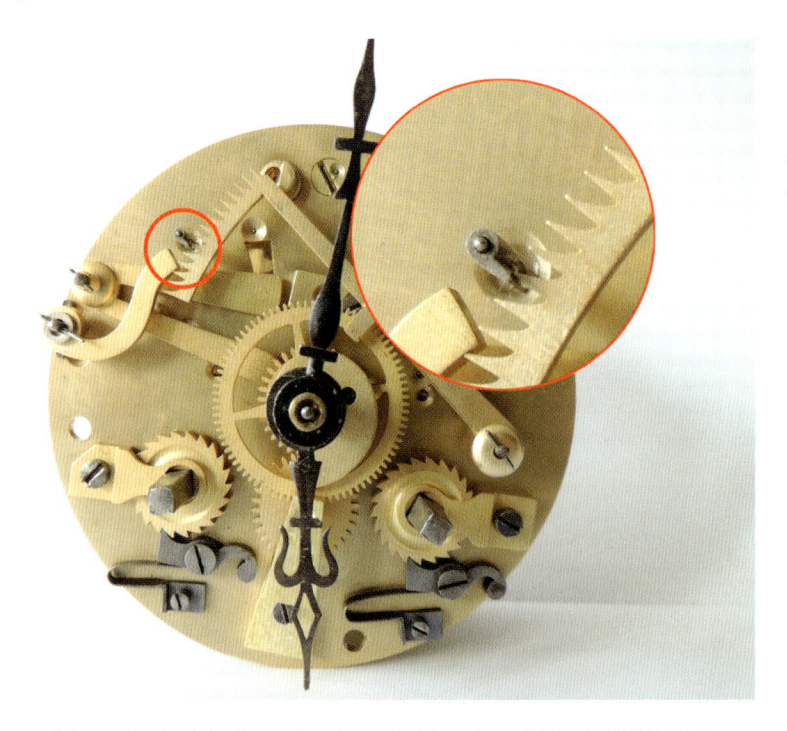

Front plate of a French rack-striking movement as the gathering pallet gathers a tooth of the rack (circled).

Lifting pin of the pinwheel raising the hammer in an English longcase clock.

arbor to control the gathering speed of the rack, and the warning wheel and fly.

The snail is screwed or riveted directly to the hour wheel (sometimes this is a friction fit so that it moves with the hour hand); as the hour wheel rotates once in twelve hours, the snail is divided into twelve steps, each of which are presented to the rack at the strike of that particular hour.

To release the strike there is a lever called the lifter; a pin on the minute wheel of the motion works lifts this lever and allows it to drop precisely on the hour. In turn, the lifter raises the rack-hook which allows the rack to drop onto the snail, releasing the gathering pallet which was locked on the rack. This allows the train to warn. The lifter carries the warning detent, which intersects the pin of the warning wheel and temporarily halts the train. On the hour, when the lifter drops from the pin, the detent drops free of the warning wheel, the rack-hook rests on the teeth of the rack, and the gathering pallet begins to rotate and slowly gather up teeth, the number of which was determined by the snail.

The train starts at the bottom with the great-wheel, powered directly by a weight or spring; this drives the second wheel pinion which, in most cases, is just an intermediate wheel with the role of gearing up the train, but in some cases (generally English clocks) there is no second wheel and you continue directly to the pinwheel. The pinwheel lifts the hammer in the same manner as a countwheel clock, whilst driving the gathering pallet arbor whose wheel rotates once for every pin on the pinwheel, or once for every blow of the hammer. The gathering pallet arbor protrudes through the front plate and carries the gathering pallet, which therefore rotates once per blow of the hammer, gathering up a single tooth of the rack with each rotation.

The gathering pallet wheel then drives the warning wheel arbor which carries the warning wheel and pin. The warning wheel then drives the fly arbor, which regulates the speed of rotation.

English longcase clock demonstrating the gathering pallet locking the strike train as its tail rests upon the rack.

Locking of the train occurs when the tail of the gathering pallet is obstructed by a pin protruding from the rack; this pin slowly works its way closer to the gathering pallet as the rack is gathered, until finally it halts the train.

TIMING MARKS

It is common in some clocks, especially in French movements, that timing marks are added to the trains during manufacture to help the clockmaker with assembly.

These timing marks are usually found as:

- a filed slash on one face of the square on which the countwheel mounts, with a matching slash on the countwheel
- a chamfered edge filed on the corner of one leaf of the hoop wheel's pinion which corresponds with a small dot on a tooth gap of the pinwheel
- small punch marks that indicate where the cannon pinion and minute wheel mesh, and, with the wheels in this position, a punch mark on the hour wheel which meshes with the minute pinion (this arrangement being more often found in the motion works of French movements).

All you then have to set up manually is the warning wheel, which should have its pin a half-turn from the detent on the lifter.

Chapter 6

Chiming Clocks

A chiming clock has a third train of gears which runs every quarter-hour, driving the chime barrel to lift and release the hammers to play a melody on a series of gongs.

The majority of chiming clocks are 'Westminster' chimers, while some allow a change of tune to include 'Whittington' and 'St Michaels'. Other quarter-chimers such as ting-tang clocks will be ignored here as they are essentially a slight complication of the strike train and can be considered as such during repair, even by the amateur.

Here we will focus on the common Westminster chime movement on which all others are built, and although it is far more complicated than I would recommend for a true beginner, it is where most will start off as they are such affordable and readily available clocks.

As previously stated, these clocks have a third train of gears with their own power source. The chiming mechanism is similar to a simple music box, driving hammers onto gongs in an order defined by the chime barrel. The chime barrel is geared to the train via an adjustable wheel, usually

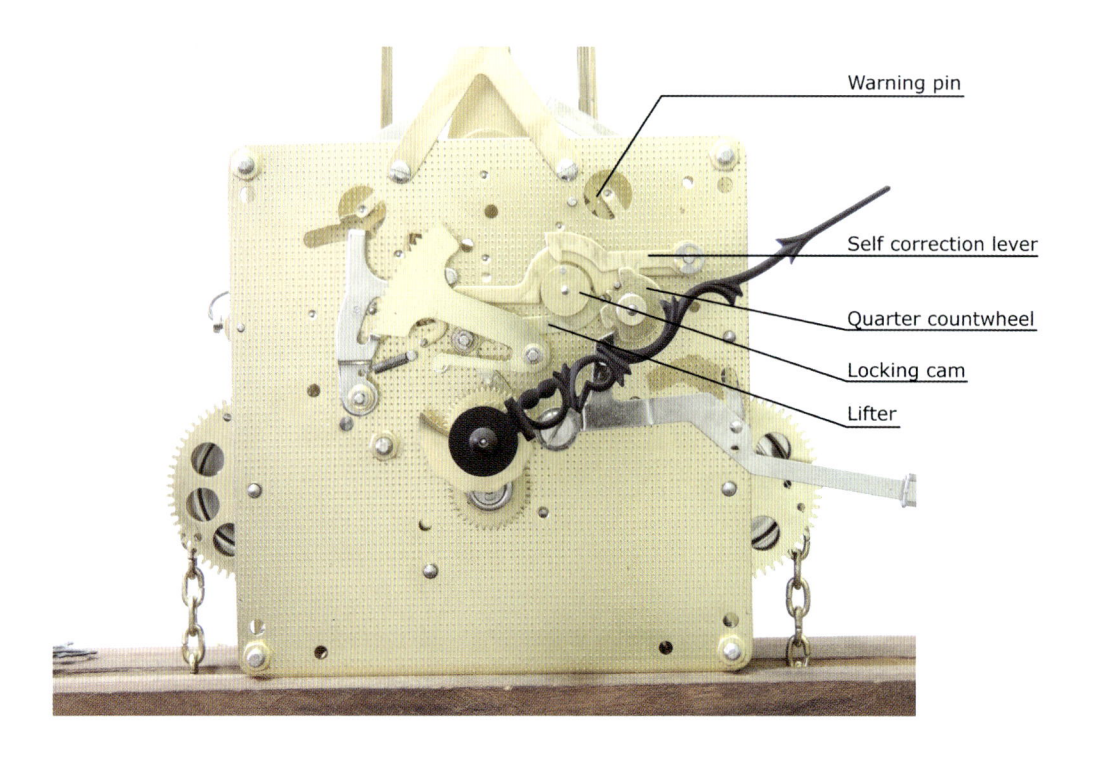

Labelled front plate of a modern chiming clock.

on the back plate of the clock. The chime barrel makes two full rotations per hour, striking four notes at quarter past, eight at half past, twelve at quarter to and sixteen on the hour, which is then followed by the standard strike of the hour governed by the strike train. Both the first four notes at quarter past and the final four notes of quarter to will be a 'straight run' of descending notes, which can be seen on the chime barrel and used as a guide during reassembly.

The quarter-chiming mechanism comes in several designs; they can have a quarter rack or a countwheel to control the length of the chime and define the quarter, and the major components can be either between the plates or on the front plate. The most common layout of these movements is a countwheel with the locking components on the front plate, so this is where we will focus our attention.

On the front plate there is a seemingly complicated series of levers and cam wheels. One cam is the quarter countwheel which controls how long the chime barrel is allowed to run, and the other is a locking wheel which locks the train and controls the self-correction mechanism.

Most Westminster chiming clocks have an automatic correction method in place to ensure that the quarters will always chime in accordance with the hand position, even if the hands are wound on without pausing for the chime train to catch up. These self-correction mechanisms can be either between the plates or on the front plate, but we will focus on the more common front plate design, where the mechanism is built into the levers and works without adjustment.

Whilst there are many designs of quarter-chiming movement, understanding this most basic and common model should allow you to adjust the theory to suit your needs. As with all things in horology, we cannot define a 'rule' and must do our best to generalize.

CHIME TRAIN

Starting at the bottom of the gear train you have the power source. This is usually either a weight-driven greatwheel with chain, or a spring barrel. This greatwheel drives the pinion

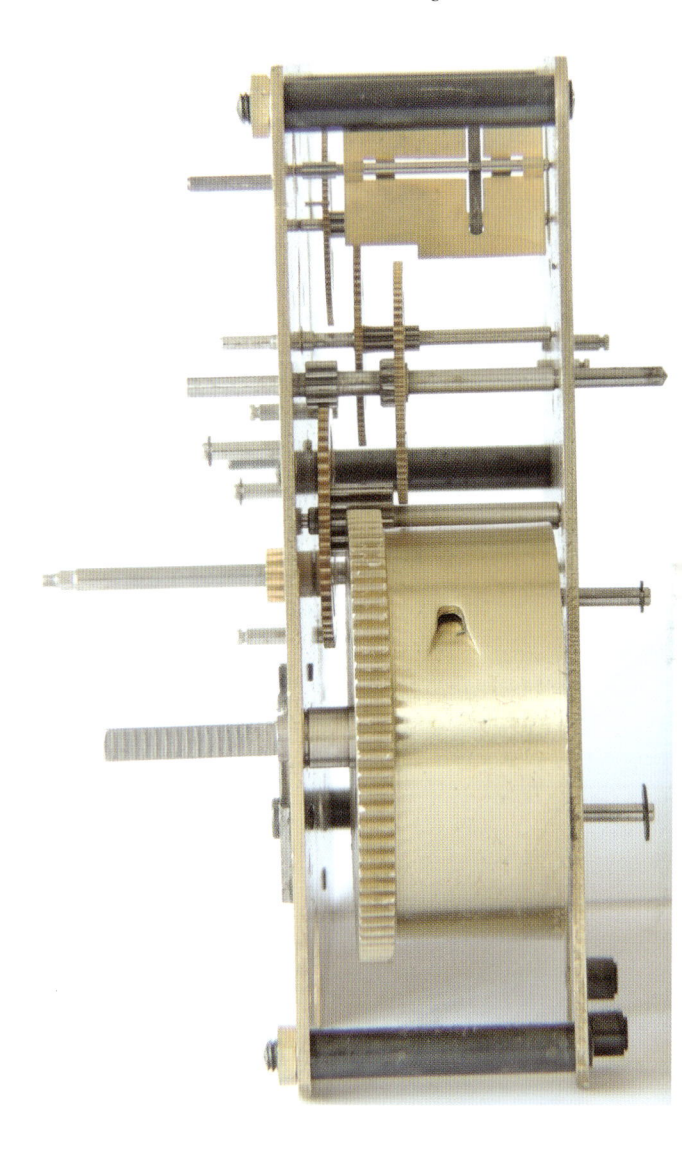

Side view of the modern chiming clock's chime train; note the extended arbors to both the front (left) and rear (right) plates.

of the intermediate second arbor which drives the second wheel. The second wheel drives an arbor which extends through both the front and back plates of the movement. This unusual arbor drives the countwheel on the front plate, and the rear extension provides drive to the chime barrel via a small run of gears from its own adjustable gear. This arbor also carries a wheel to drive the next arbor in the train; this is the locking arbor which provides an extension through the front plate onto which the locking cam is attached. Beyond this locking arbor is the usual warning wheel and fly.

These movements can be put together without any special attention to the relevant positions of the wheels and arbors, because all functional components, locking cam, countwheel, etc., are external and can be adjusted. All that must be taken into account is the amount of warning provided. Similar to the warning of a strike train, chime trains require a small amount of warning to allow the locking detents to fully release, but not so much that the hammers of the chime barrel begin to lift. Chime trains require a quarter-turn of warning, no more. This quarter-turn of warning nearly always points the warning pin toward the fly, and the pin is often fitted at the end of one of the crossings of the wheel; this allows you to hold it in place with tweezers as you set up the front plate of the train.

FRONT PLATE

The front plate of the train consists of a series of levers, including the warning detent which is lifted indirectly by the cannon pinion at each quarter. As the lever is lifted it simultaneously blocks the path of the warning pin and raises the locking detent. The locking detent rises to release the locking cam, therefore releasing the entire train until it is held up by the warning detent.

The self-correction mechanism is a small lever which pivots on the same centre as the locking lever, but is free to rise and fall dependent on the position of the quarter countwheel or locking plate which has a small cam on its underside (nearest the plate). This cam has a cut-out into which the re-synchronization lever will fall just before the fourth quarter on the countwheel. With the self-correction mechanism in this position it will hold the locking cam in place regardless of what the locking detent is up to. When the hour is reached, the cannon pinion provides a taller lifting pin, which will raise this self-correction mechanism out of the cam wheel as it lifts the locking detent, freeing the train to chime the final quarter.

As the final quarter begins to chime, the countwheel often has a small bump on the cam which actuates via a lever and releases the strike train, holding it in the warned position until the chiming is complete.

Levers of the modern chiming clock movement in the fully locked and self-correcting position.

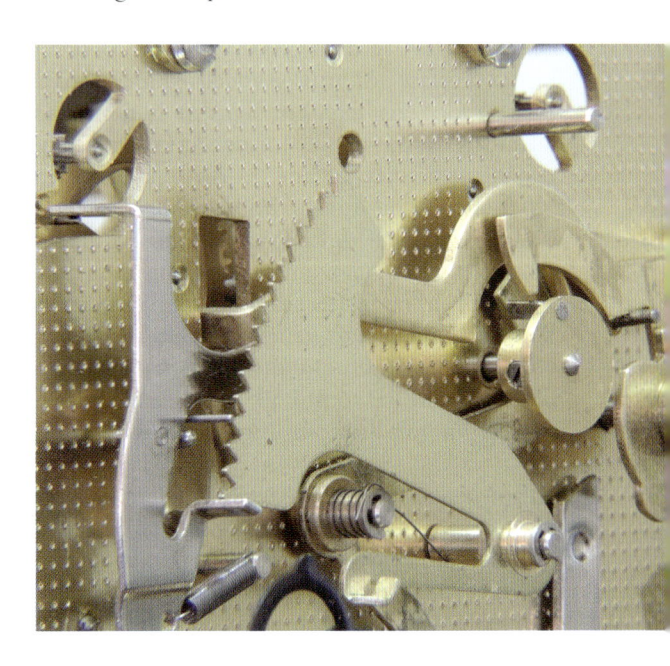

Levers of the modern chiming clock movement unlocked to allow the fourth quarter to chime.

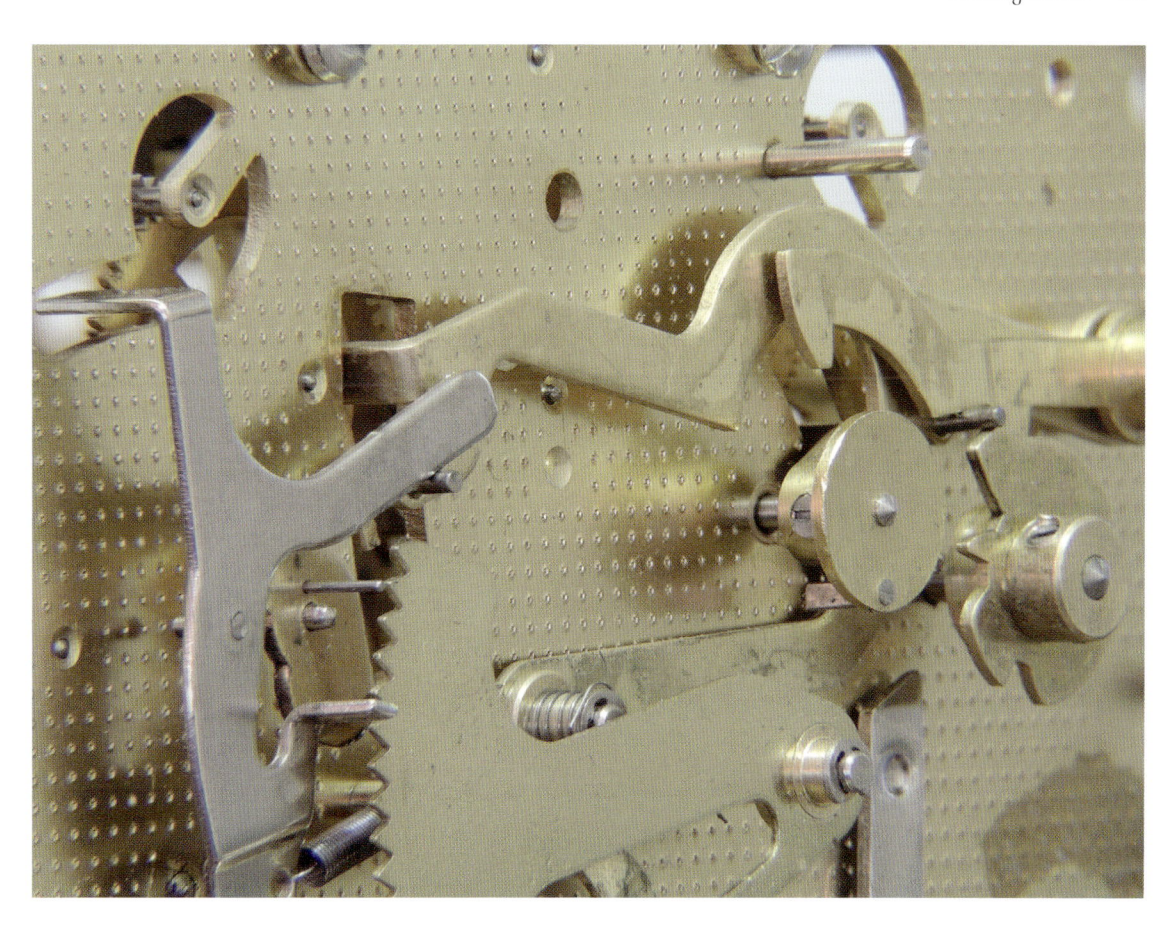

Levers of the modern chiming clock movement as the quarter countwheel unlocks the strike train.

REAR PLATE AND CHIME BARREL

On the rear plate you have one extended arbor, onto which is fitted a large wheel held by two grub screws. By loosening these screws you are able to rotate the chime barrel in relation to the train in order to set up the chimes properly. This wheel will drive the chime barrel, either directly or via two intermediate wheels which rotate on studs and are held in place by C-clips.

As the chime barrel rotates, the small pins lift and release the hammers in sequence. The hammers should be free or else the chime will be sluggish or fail entirely. If the clock has multiple tunes, then a lever or cam wheel will be provided to push the chime barrel to one side by a predefined amount, lining up a new set of pins with the hammers.

Chapter 7

Getting Started

Mechanical clocks are expected to run for years on end with little to no attention to the mechanism. While there is no other mechanical device which would withstand this level of abuse, most of my customers are horrified to hear that their clock is horribly worn when they finally bring it in.

Five years after a clock is serviced, what remains of the oil is dry and full of dust, rust and brass. With the constant rotation of the pivots grinding this mixture into the pivot and its hole, the result is an elongation of the pivot hole and a badly worn pivot.

Meanwhile the mainsprings are repeatedly wound up and down, every week for as many as thirty years in some cases, with nothing but dry grease for lubricant. The hooking eye is constantly under pressure against the barrel hook, while the inner coils get folded over the barrel arbor hook, forming deep impressions in the spring steel. Eventually cracks appear, and when the clock is next wound, they will propagate along the steel which will let go with a bang. As the spring breaks, it puts a tremendous side load on the teeth of the wheels and pinions, which is passed in turn to the pivots. If you are lucky, nothing will break, but in many cases you will have broken teeth, pinions and bent pivots to deal with.

Regular maintenance can help avoid all of this hassle and expense. I regularly spot cracked mainsprings before they break and I am able to correct the issue before it gets expensive. Regular maintenance means full overhaul every ten years, with fresh oil being applied every three to five years. If you do not stick to this then your clock will wear dramatically and the repair bills will go up as the value of your clock goes down.

The filthy escape-wheel rear pivot of a French movement; no wonder this clock would not run.

Severely worn pivot hole; the shoulder of the arbor is clearly visible through the wear gap.

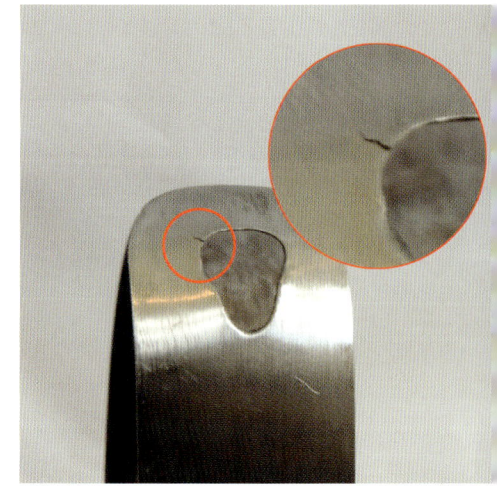

Cracked mainspring outer hooking eye, caught and corrected before catastrophic failure.

If your clock is especially valuable, and you want to keep it in the best museum-worthy condition, then it should be overhauled every four or five years to keep the oil fresh and clean and to keep the pinion leaves free of dirt.

A SYSTEMATIC PROCESS

Clock servicing and repair should be tackled in a systematic way. When you have a system, it is harder to miss steps and you are far less likely to end up with problems further down the line.

1. Inspection of the complete clock.
2. Disassembly and inspection of the movement.
3. Cleaning the movement.
4. Repairs and component manufacture.
5. Cleaning again to remove fingerprints.
6. Oiling and reassembly.
7. Testing the movement out of the case.
8. Final testing and set-up of the movement in the case, on a level surface.

Each of these steps has its own internal steps to be followed, details of which are provided throughout this section. In the resources section of this book a checklist has been provided which can be photocopied and placed on the workbench.

RECORD KEEPING

A professional will be expected to keep a record of repairs and it is a good practice to get into for anyone repairing clocks.

My record system starts with a large tag which is tied to the clock on arrival. One section of the tag is cut away and given to the customer: this gives them a record of where their clock is being serviced, as well as providing a ticket onto which we can print our terms and conditions.

The middle section is a standard 15 × 10cm (6 × 4in) index card onto which we can write the customer's details and, on its rear, the repair record, which is laid out to make keeping good records simple.

Finally, at the top of the card is a small ticket which remains tied to the clock throughout its stay. This ticket stops the clock from becoming separated from its movement (which is kept in a box with the index card during repair). On its rear is a test record where we can note the date we wound and set the clock for testing, and the result at the end of the test period. Several test periods are provided for problematic clocks.

An exact copy of the record card I use can be found in the resources section at the rear of the book.

DISASSEMBLY

Basic Checks

Before beginning work on any clock it is a good idea to run a few basic checks. This is done for safety, to avoid damage to oneself and to avoid further damage to the clock, but mainly to discern why the clock is not working.

When you approach a stopped clock it is best to take a minute to look without touching, familiarize yourself with its type, decide whether or not it is likely to have a pendulum or weights which you will need to remove, and to run a mental risk assessment. (There is nothing worse than rushing in to collect a clock, only to find that the handle is loose or for the rear door to fall off.) Check that the hands are not touching each other or the dial and that the pendulum is free to swing without obstruction. I do this regularly when I collect a clock for the first time, and it can be done discreetly while making small talk. If all is well, try to start the clock. If there is no tick check that it is wound. It is now time to expose the movement and take a closer look.

While you are studying the clock it is best to ask the owner a few questions about its history: When was it last seen to professionally? Has it had any persistent problems? How has its timekeeping been? These can all suggest further problems which you will need to assess before accurately quoting for repair. It is false economy to part-repair a clock which has not been overhauled in fifteen years; it will stop in the near future and, as the most recent known repairer, the owner will hold

The second hand of this longcase clock is resting against the hour hand, stopping the escape wheel from turning.

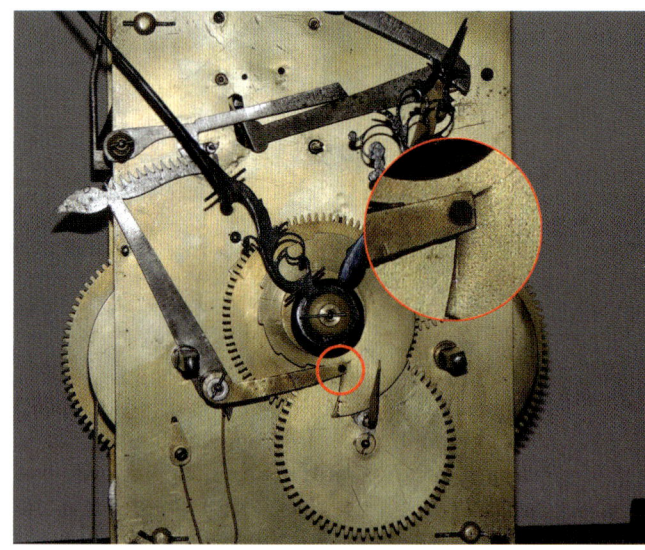

The rack-tail of this longcase clock is jammed against the step of the snail, causing the clock to stop at ten to one.

you responsible. It is better to have a reputation for being expensive for good work than for being cheap for unreliable work.

On some clocks you can expose the movement simply by removing the hood or a rear or side panel (but check first that the winding key is not balanced on top of the hood). Watch the escape wheel while leading the pallets to and fro by hand (for clocks with a balance wheel you cannot do this on site). You may find that the clock is wildly out of beat, or that the strike has failed and that by turning the hands backwards slightly, you can release it. (Note that it is common for longcase clocks to stop at around twenty to one if the owner goes away and the clock is not wound: as the strike runs down the rack-tail can jam against the step of the snail.) Check for obvious signs of wear; little piles of brass filings and rust are common on the seat boards of poorly maintained longcase clocks. If there is any sign of wear or the movement is filthy or covered in grease, then it will need to be taken back to the workshop. If you can free the escape wheel and set the clock in beat and it is showing no obvious signs of wear or grime, it would be wise to set it up and watch it closely for five minutes.

Safe Transportation

Transporting the clock safely is another issue which must be dealt with before you can get it to the workshop. Firstly (except for longcase and Vienna-type clocks), the pendulum must be removed: if this is not done it will crash around causing damage to the suspension, the crutch and possibly the escape teeth and pallets, as well as scratching the inside of the case. Any weights need to come off too, even fake decorative weights which are sometimes used on reproduction spring-driven clocks.

Large packing crates full of bubble wrap are essential; I leave a spare in my car and take one to every job. They stack up neatly when not in use and reassure the customer that you will look after their clock. If you are transporting cases regularly, I would suggest buying a few professional removal blankets.

For larger clocks, you will probably want to leave the case behind unless it also needs to be restored. If it is simple to do so, I would suggest this with all longcase clocks, Vienna regulators, large slate clocks and dial clocks which have had their cases screwed firmly to the wall. This book will provide plans for making test rigs which will allow thorough testing of these

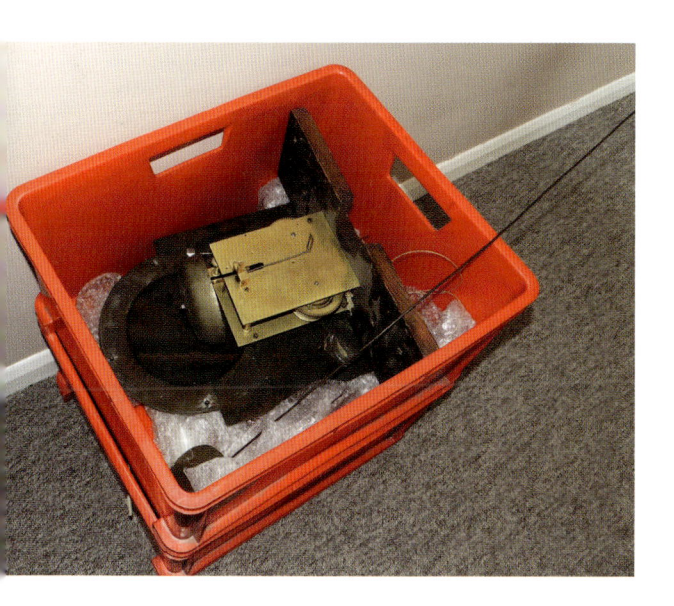

Getting Started

Packing crate lined with bubble wrap for safe transport of a longcase clock movement; such crates can be stacked two high if appropriate.

Hood latch of a longcase clock as seen from the inside of the case; do not forget to undo this before tugging at the hood.

styles of clock without having to store the large cases in your workshop. Remember, if leaving the case behind, bring away the seat board for longcase clocks, the mounting bracket for Vienna regulators, and the dial and surround for dial clocks.

For longcase clocks you will need to remove the hood first, open the case door and reach up towards the front of the hood; sometimes there is a latch which should be checked for at this stage. This will usually be left of centre as you face the clock, and is usually a simple wooden block which pivots, and sometimes it is a bolt lock. It is more common for there not to be a latch, but if there is one, and you miss it, tugging the hood may cause the entire case to fall. After releasing any latches, the hood will usually slide forward, although some designs rise upwards above the movement. You may need to brace the body of the case as sometimes the hood can be a tight fit. Never pull on turned pillars or any ornamentation on the hood; the glue is old and they come off easily.

Unhook the weights, but be careful not to set them on a tiled floor; if they slip from your hands you will have a very unhappy customer. I keep a box of bubble wrap nearby to place them in. On another note, it is well worth advising

The hood (highlighted in red) will slide forward as a complete unit allowing full access to the movement inside.

a customer with a tiled floor to stuff the bottom of the case with a sandbag or cushions, as a single slip of the weight during winding could become a very costly floor repair.

Finally, the pendulum must come off, but before its removal the seat board must be checked for screws. A longcase movement and dial are very front-heavy and will fall forwards once all weights and the pendulum are removed. If the seat board is not firmly fixed down, with the movement fixed to it, then keep one hand on the front of the dial as you release the pendulum, then lower it into the case and set it down while you lift out the movement. After the movement is safely packaged in bubble wrap (either face down or standing up, never on its back), then you can retrieve the pendulum.

If possible, I would advise that you unscrew the suspension block from the top of the pendulum rod although sometimes they are soldered if the thread has worn.

Vienna regulators are easier. Unhook the weights and bubble wrap them, leaving the pendulum until the movement is out of the case. To take the movement out, undo the two knurled bolts which go into the sides of the seat board (behind the dial, roughly positioned by the 4 and the 8) and slide the dial and movement forwards. Place this face down in bubble wrap. You can now unhook the pendulum from its suspension. In order to test the movement out of the case, you need to take the bracket that the seat board and pendulum hang from: this is usually five screws; put the screws back in their holes so they cannot be lost, as you will be using your own in the workshop.

Two knurled bolts (highlighted in red) will release the movement from the case when fully removed.

The suspension blocks of longcase clocks will usually unscrew from the rod.

Do not forget the mounting bracket; it is held in place by a number of wood screws (highlighted in red).

French clocks with large slate cases are easier still. Open the rear door and remove the pendulum (you may need to remove the bell for access; both of these should be transported in a padded envelope and its nut replaced on the movement). Release the two screws on the door frame, one on each side; keep one hand on the front door as you do this so the dial and movement do not fall out. As you remove the last screw you will need to hold the rear door in position to stop it from falling. I like to pop the screws back through the holes in the door frame and close the door; this way they cannot be separated. Leave the rear door and the case behind as you do not need them for testing.

Most French movements can be released from their case by removing these two screws (highlighted in red).

Dial pegs (highlighted in red) are used to hold most fusee dial clocks together; pull them out to release the entire bezel and movement assembly.

Finally, the easiest of them all; if a dial clock has been firmly screwed to the wall, I prefer to leave the case behind: it makes for a quicker visit and stops any chance of plaster flaking off the walls as you unscrew the clock (your responsibility to put right in the customer's eyes). Reach up inside the bottom door and unhook the pendulum; there is usually also a side door if you need to use a second hand. Remove the four dial pegs, keeping in mind that the dial will fall forwards once all the pegs are out. Lift off the dial, movement and bezel all together. Place the four pegs back in the holes on the case so they do not get lost, put the dial face down in bubble wrap and wrap the pendulum.

Putting the packing crate in the back of the car is fine. I like to make sure that nothing can slide around and I accomplish this simply by having two packing crates to fill the rest of the boot space. Smaller clocks in their cases which are too tall for a crate are stood on the passenger seat wrapped in a blanket with the seatbelt on. This way I know it will not move and I can adjust its position if I need to. Carriage clocks are best off being well wrapped and placed in a crate full of bubble wrap.

In the Workshop

Once the clock is in the workshop, and you have the workbench cleared and ready to use, you should first make the clock safe to handle. To do this you need to remove the power from the mainspring; the usual method for doing this involves removing the movement from the case and dial first.

The process for removing the movement from the case varies from clock to clock so a bit of common sense is needed to read between the lines. We have already done this step for four types of clock and can skip to removing the dial and hands for longcase clocks, Vienna regulators, French slate clocks and dial clocks.

English bracket clocks often have a bracket on each side of the rear of the movement, and a screw on each side of the case towards the rear; simply remove these screws, and pull the movement and dial from the rear of the case.

Most post-1900 mantel clocks require that the hands are removed first, because as the

English bracket clocks often have two simple brackets (highlighted in red); when they are removed the movement will pull out from the rear of the case.

movement is removed, the dial stays attached to, or is part of, the case.

To remove the hands you either release the taper pin by pressing it through the centre arbor with smooth-jawed pliers, or undo the hand nut which requires holding the minute hand steady as you loosen it. The hand washer can then be lifted off, followed by the minute hand, which is usually a loose-to-medium fit on the square of the cannon pinion. The hour hand is sometimes secured by a screw or cross pins so check for this first, otherwise they simply pull off. Hour hands can be quite tight; if you cannot remove one with your fingers you may need to grip the hand centre with a blunt pair of end cutters (being very careful of the dial) and pull straight out, Applying some oil to the centre sometimes helps. If this fails, special hand-removing tools can be bought or made.

There are several designs of hand removers available to purchase, for smaller hands I use what is commonly known as the presto type of hand remover, these pull the hands straight up, whilst using nylon pads to protect the dial, and for larger hands, I use a home-made lever type hand remover. These have the appearance of a pair of bent screwdriver blades, bent no more than 30 degrees, and are used on opposite sides of the hands to even out the pressures which may otherwise bend the arbor.

Do not be tempted to lever against the dial with a screwdriver. Second hands usually pull off in the same fashion and again can be extremely tight; there is usually no clearance for getting a tool under and pulling here so hand removers may be needed. If you can get behind the dial, you may be able to release a stiff hand from this side of the dial, which is much safer in case of slippage. Auxiliary hands usually do not need to be removed at this stage as they operate levers and cams often attached to the rear of the dial, but if you cannot see in from the side of the case to check, remove them just to be safe.

With the hands off, you can remove the movement. In modern mantel clocks this is usually done by reaching into the case from the rear with a long screwdriver to release the

Removing clock hands is usually a case of pressing out the taper pin.

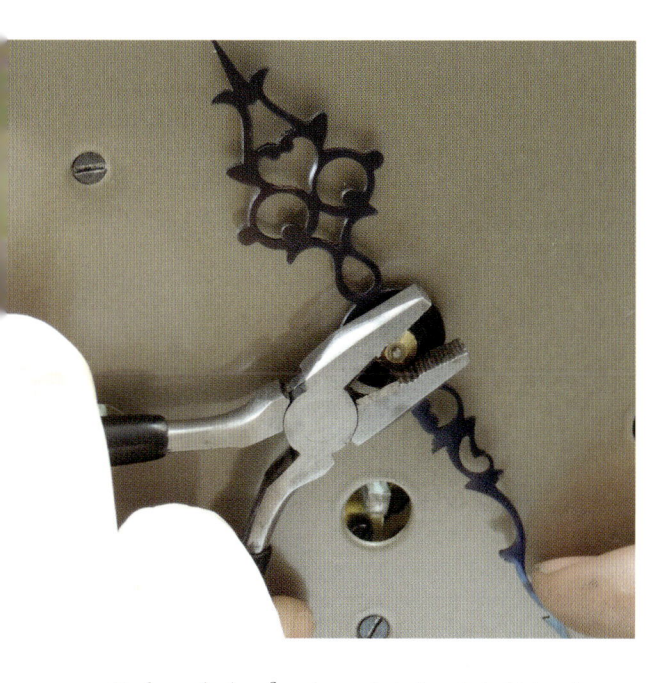

Modern clocks often have their hands held in place with a knurled collet or nut; they can be released by supporting the minute hand to stop it rotating backwards, and unscrewing the hand collet.

Second hands are often very stiff; note the use of rubber pads to protect the dial from damage.

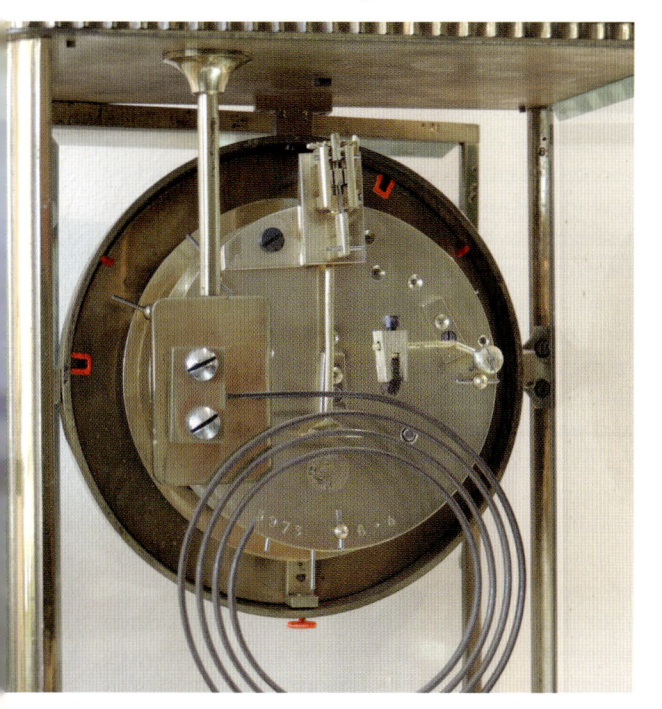

French four-glass clocks are held in place by two pins and a screw (highlighted in red); release the screw to allow the movement to rotate in the bezel.

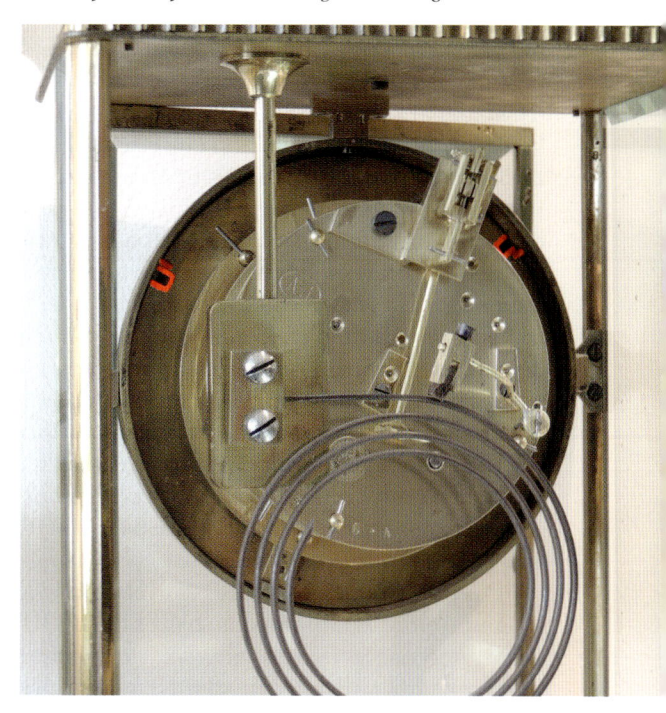

Rotate the entire movement and dial until the pins pass through the cut-outs, then manoeuvre the movement out of the case.

wood screws in the movement corner brackets. This is best done with the clock face down in your lap. When the screws are free you can lift the movement out of the case and set it to one side; the hammers often make this a bit of a fiddle, but it will come out. At this point I would suggest you put the screws back into the case and close the door. If these screws go missing it is very important to replace them with screws of the correct length, otherwise they will burst through the front of the case, damage which I have seen from previous repairs far too often.

Elliott mantel clocks often require removal of the brass dial from the front of the case to gain access to the four screws which release the movement.

Modern wall clock movements are often screwed to the back board of the clock from the outside rear of the case. In this situation it is best to stand the case on its side in your lap, support the movement with one hand and remove the screws or nuts with a tool in the other hand. When the final screw is out the movement and dial will try to fall.

French four-glass clocks are often fitted in a very simple but clever way. The dial surround stays attached to the case while the dial comes out with the movement. At the rear of the dial, behind the numeral 6, is a screw; slacken it so the dial will rotate in the dial surround. At the top of the dial surround are two pins, and in the dial are two cut-outs; rotate the dial until the pins pass through the cut-outs so the dial and movement can be lifted out.

Clocks with a seat board can have their movements removed in one of two ways. You can release the screws which hold the seat board in the case and slide the seat board out. Alternatively, you can remove the two screws that hold the movement to the seat board and lift the movement away on its own. Sometimes there will be a bracket at the top to support the movement; usually this bracket slips onto a fixed bolt in the case and is held in place with a nut.

There are so many ways of fixing a case that I cannot cover all of them; I am sure that there are still at least several common designs which I have not encountered myself. That being said, knowing the common methods described so far will help with anything unusual you may encounter.

Next you remove the dial, so you can gain access to the ratchet wheel and click. It is possible to reach these components with the dial on, but a much safer grip can be had on the components with it removed. On carriage clocks the ratchet and click are on the back plate of the clock, but I recommend removing the dial first to reduce the risk of damage. I always recommend removing the dial and hands at the earliest opportunity.

Dials can be held on in a number of ways but most often they are pinned with taper pins. To remove them simply press out the taper pin from the small-diameter end, or use pliers or cutters to pull from the large-diameter end of the pin. Try to avoid levering off from the edge of the plate, which causes damage.

Other methods used to hold dials are small pivoting clips which are riveted to the front plate of the movement; they simply swing around into a slot in the dial foot and hold the dial in place. Separate clips are commonly used in modern clocks; they are made from spring steel and are shaped to hold the dial foot in place with spring pressure.

Some clock dials have what is known as a false plate. This is a plate onto which the dial is pinned, and the plate itself is then pinned to the movement. There is generally no reason to remove the dial from the false plate, unless you intend to polish it or repair damage, so when removing a dial with a false plate, just unpin the false plate from the movement.

Now that the movement is out of the case, the hands and dial are removed and safely tucked away, it is time to release the power from the main springs. The safest way to do this is to use a let-down tool. This is a screwdriver-like tool with interchangeable ends of the twelve common sizes of winding key. Fit the end to the handle which best suits your needs and place it on the winding square. Grip the handle firmly and put a small amount of winding pressure on the winding square to free the click, which can now be released with the other hand. Allow the handle to slip slowly in your grip until the spring has no power left. To do this, I find it easier to grip the movement between my knees while seated; the movement must be clamped otherwise it will spin as the spring is released.

Sometimes you have to remove tight pins by levering against the plates; if you do, use rubber pads to protect the clock from damage.

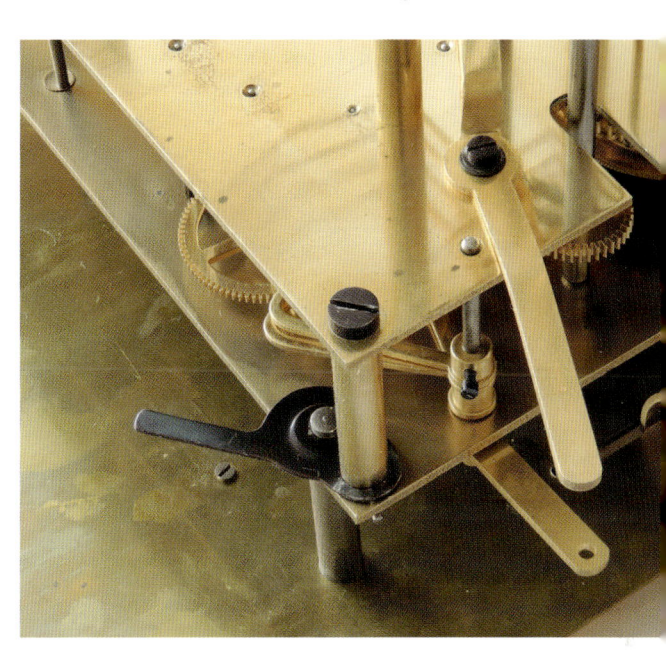

Some dials are held in place by swivelling clips; to remove the dial simply rotate the clips away from the dial feet.

To release the power from a mainspring use the proper let-down tool; release the click and unwind the spring carefully.

Fusee clicks are sometimes accessible through a small hole drilled between two teeth; this method is preferable to letting the train run down.

When the power is off completely, unscrew or unpin the ratchet wheel and remove it. Repeat this process for all squares.

For removing the power from fusee clocks this method will not work. The large ratchet on the front plate is for 'set-up' power and is often larger than your let-down tool, and the ratchet and click are housed within the fusee greatwheel. Some clocks have a small hole drilled in a tooth gap through which you can insert a pin to release the click (while gripping the winding square in the let-down tool) but this is not common. It is best to oil the pivots and remove the anchor; the clock will run down until the spring is almost unwound. When the wheels stop turning, there is only the set-up power in the barrel left to release. To do this, screw a large hand vice or T-chuck onto the barrel arbor square, grip it tight and unscrew the large click. The set-up power is usually half a turn or so; release it with control and place the click back in the ratchet teeth as you pause to re-grip the vice. When the power is off I like to cut the old gut-line or unhook the chain before I continue inspecting the movement.

Platform Escapements

Working on platform escapements takes a gentle touch. The slightest slip can result in broken pivots and jewel-holes, while too tight a grip with the tweezers can send tiny screws shooting across the room, never to be seen again. Before starting work on a platform escapement, I highly recommend that you take some time to sharpen your screwdrivers and tweezers, sweep the floor around you and tidy the workbench to aid in finding any dropped parts. I would also recommend working in an apron or with a tea towel draped across your lap to catch parts. I have a small table which I place on my workbench when working on platform escapements. Its purpose is to raise the height of the work surface to make small parts easier to see, to provide a fence which should catch parts as they fire out of my tweezers, and to give me a place to store my best tweezers and screwdrivers, for use only on the finest work. Plans for this table can be found with all other plans in the resources section of this book.

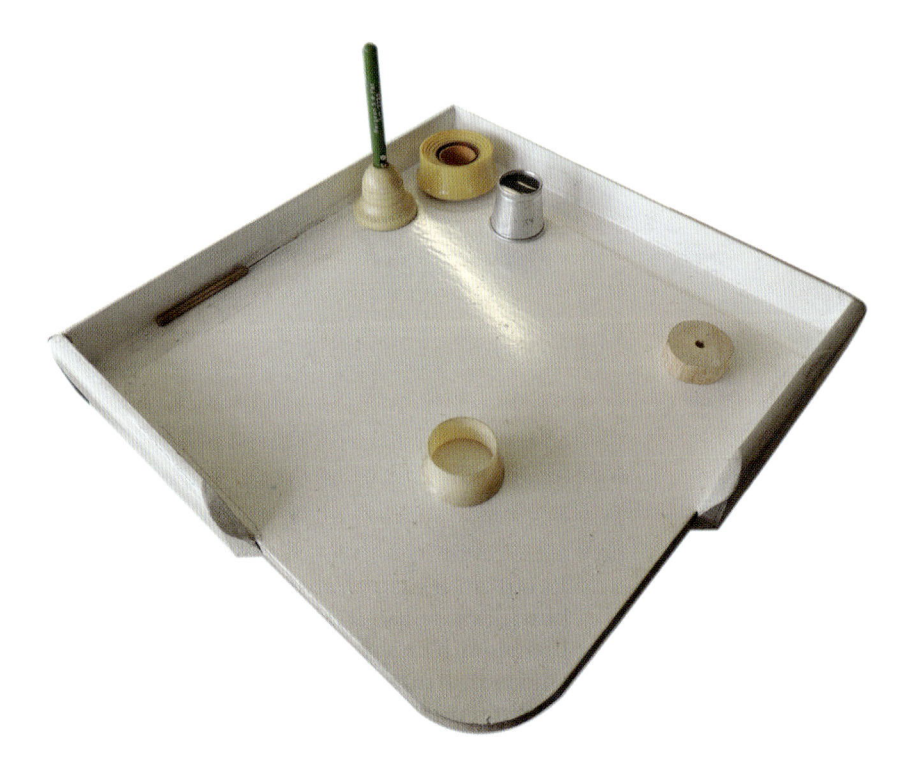

I use this small table for working on platform escapements; the fence catches any parts which get dropped and the white finish makes them easy to spot.

Platform escapements are usually attached to the movement with a screw in each corner (highlighted in red).

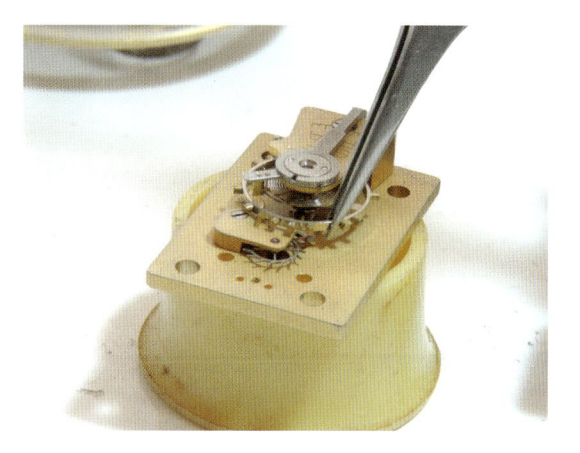

Test the end-shake and side-shake of the balance with tweezers before disassembly so that problems can be spotted early.

With the power removed you next remove the platform escapement if the clock has one. This is so easily damaged that it would be foolish to start work on the movement with it in place. They are usually held in place by a screw in each corner; release these screws and lift the platform away, being careful to work the escape wheel lower cock around the contrate wheel without damaging it. Place the platform on a boxwood ring on the small parts table for disassembly.

Inspect the balance staff or cylinder by gripping the balance rim with tweezers, and test for end-shake and side-shake. Be gentle, as the slightest slip will cause damage. Broken pivots will be immediately obvious as the balance wheel will flop around. A broken cylinder is even more obvious, as the balance will not sit square and parallel to the platform. Excessive or insufficient end-shake should be noted and the reason for it found. Inspect the end stones of the balance; if they are broken, missing or incorrect in size, they will need replacing. Remember what was learnt in the timepiece chapter about how the pivots sit on the end stones; the tapered portion should not be able to pass into the jewel-holes, and the straight portion should be well supported.

The interaction of the index and hairspring is important. Check that the index travels along the curve of the hairspring without distorting it as the index is moved to and fro. If not, make a note to adjust the hairspring during assembly; while you are at it, check that the hairspring sits flat, and that it forms concentric circles around the balance staff.

Finally, check the interaction between the roller and lever, or cylinder and escape wheel. The impulse jewel should pass freely into the lever with the same amount of freedom about the guard pin on each side of the line of centres. A refresher of platform theory here is a good idea. A cylinder platform should interact properly with the escape wheel; if the interaction is too shallow, the impulse faces of the teeth will drop onto the impulse faces of the cylinder. Make notes about the adjustments to be made, and also check that the cylinder does not contact the arm of the escape wheel at either extreme of its end-shake.

Move the index over to one side so that you can access the screw in the centre and release the screw(s) holding the balance cock in place. Lift the balance cock away, and if it is tight (raising a tight balance cock can be difficult, and it is important that you do not allow it to tip forwards crushing the balance staff or cylinder as you prise it off), try using a screwdriver in the relief slot on the back of the cock and give it a twist to release it. Alternatively, turn the platform upside down and use the point of the tweezers to push the steady pins from underneath. Put the screw in the cleaning basket and lift the balance cock, complete with balance wheel, to one side, letting the balance hang, and turn it upside down so that the balance rests back in its jewel-holes. Leave it to one side for later.

Using a loupe, study the interaction between the escape wheel and lever by applying power to the escape wheel, while slowly moving the lever back and forth. Observe the drops, run to banking, etc., as detailed previously. Skip this step with a cylinder escapement; instead, spin the escape wheel and check that it runs true.

CLOCKWISE FROM ABOVE:

With the index moved over, the balance cock screw can be accessed.

A relief slot is often provided to help with the removal of a tight balance cock.

With the platform upside down and well supported, pressure can be applied on the steady pins to safely release a tight balance cock.

Place the balance assembly to one side; resting it upside down like this is safer for the pivots and hairspring.

Test the locking action of the pallets with pegwood and tweezers now so that problems can be spotted early.

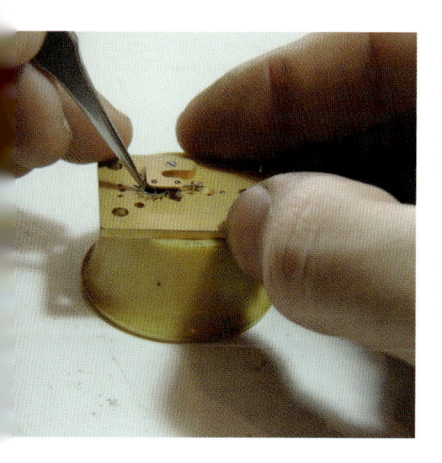

Test the end-shake and side-shake of the escape wheel by gripping the crossings with your tweezers.

Test the end-shake and side-shake of the lever by gripping the arm with your tweezers.

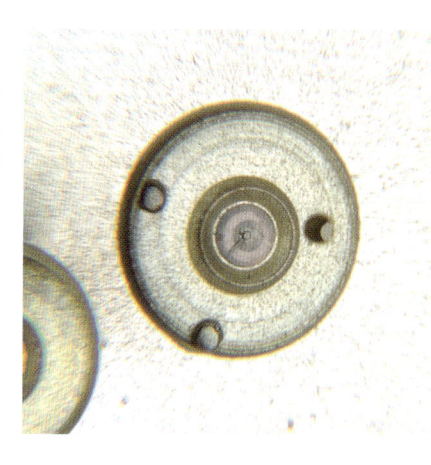

A perfect example of what to look out for; a cracked jewel hole like this will increase friction and wear to the balance pivots.

Check all pivots for excessive side- or end-shake by testing with the tweezers, giving special attention to lever pivots as excess slop here can be troublesome. What you are looking for is more movement than necessary; things need to be loose but not sloppy. Hold the escape wheel by one of its crossings with the tweezers, and pull it back and forth, side to side and up and down, then hold the pinion at the other end of the arbor and repeat the test. Jewel-holes tend not to wear, but be suspicious of brass pivot holes at this scale; wear here is easy to miss. To check the lever, hold it along the neck with the tweezers, and test it in the same manner as before. Make a note of any repairs to be made.

Undo the screws that hold the escape wheel and/or lever in place, and release the bridge (or cock) in the same way as the balance cock. Study the bridge closely, looking for distortion or burrs, and the jewel-holes for any wear, cracking or chips. Make a note for later. Next, lift out the escape wheel and check the teeth and pivots for wear, also the pinion leaves which wear frequently on dirty platform escapements. Lift out the lever and once again check the pivots, pallet stones and the notch for wear or damage. Place all the components in the basket for cleaning.

Turn the platform upside down and remove all of the end stones, taking care not to lose the tiny screws which hold them in place. Inspect for damage and place them in the cleaning basket. Finally, inspect the jewel-holes in the platform and if it is plated with tarnished silver, soak it briefly in silver-dip or wipe with a polishing cloth to remove the tarnish (which does not wash off in clock-cleaning fluid).

You can now begin to disassemble the balance cock assembly. Release the boot on the index by turning with a screwdriver in the slot. If it is tight, turn it with tweezers, and if there is a screw holding the balance stud to the balance cock, release but do not remove it. Lift the cock and turn it the right way up; rest the edge nearest the stud on a solid surface such as an anvil and, holding it in place by hand, use the tip of your tweezers to push the stud through and release the balance wheel. Sometimes a stud is a bit stiff and will benefit from being loosened by being twisted back and forth with the tweezers before pressing it out. Study the balance rim, balance staff and hairspring for damage or wear, or inspect the cylinder for wear. The cylinder escape wheel has a tendency to cut into the cylinder at the point on which it rests; this is visible as a groove in the cylinder surface, and if it is bad you must find a way to repair or avoid the wear groove. Check that the impulse jewel of a lever escapement is not loose or chipped, and that it is upright in its setting. Place the fully assembled balance in its own compartment of the cleaning basket. There is no need to remove the hairspring.

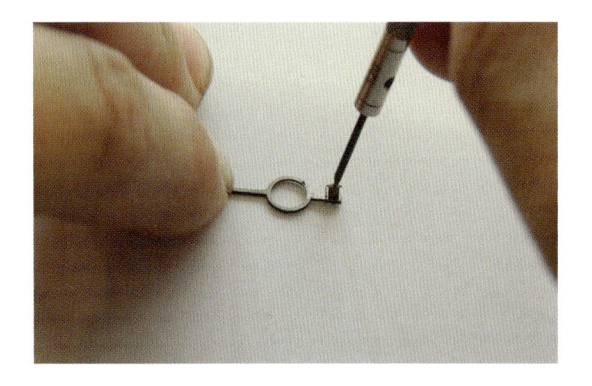

The index boot can be released by turning it with a screwdriver in the slot provided.

Releasing the hairspring stud from the cock is easily done by pressing it out with your tweezers.

Release the two screws holding the end stone in place and the rest of the balance cock assembly will come apart; inspect the parts and put them in the basket.

There are so many platform designs that I cannot cover every eventuality. They are generally similar although some disassemble in a different order. It is common, however, to come across modern platforms with Incabloc shockproof jewelling; these require you to manipulate the delicate shock spring in order to release the end stone and jewel-holes for both the top and bottom of the balance wheel. Use a sharp piece of pegwood to release one arm of the shock spring at a time. When open, it will hinge up to release the jewel-holes, which can be removed with a piece of Rodico rolled to a point; then, with the tweezers, lift away the setting from the Rodico. The end stone will stay stuck to it. It is now fully disassembled and can be cleaned.

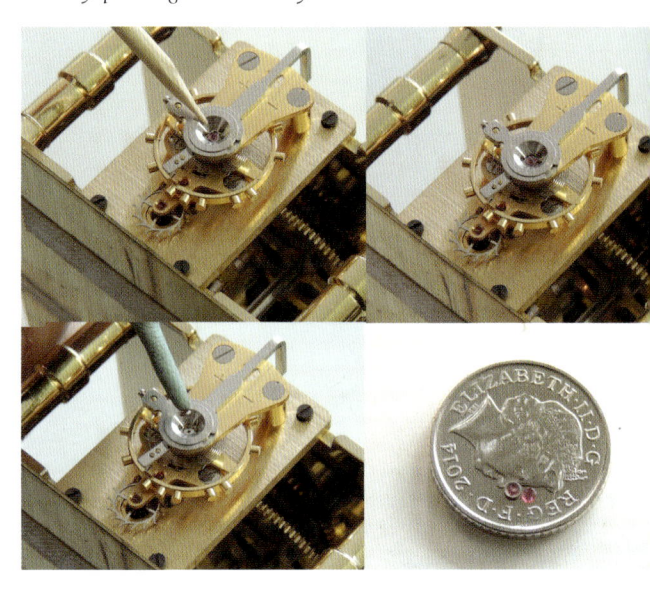

Stripping the Plates

With the power totally removed from the springs you can begin to work on the movement. But before you do that you should have a good look and familiarize yourself with its layout; look at the screws and take note of where short screws are used to provide clearance, study the strike or chime mechanisms and make note of where the warning pins sit when the train is locked. Taking pictures or making a simple sketch is advisable. Apply power to the trains by putting turning pressure on the second wheel with your fingers and watch it strike and chime; check that the

Modern shock assemblies can be disassembled with pegwood and Rodico.

Create a hollow in a piece of pegwood to release these twist-lock shock springs by pressing down and turning.

motion is not jerky and does not jam which could indicate a bent tooth or pivot. Apply some power to the third wheel and watch the escapement as you lead the crutch to and fro. The escapement's action should be the same on each side, the same amount of drop onto each pallet, the same amount of impulse. You should be able to see the escape wheel rotate during the impulse phase of the cycle; if not, then it is resting rather than impulsing and the pallet angles are wrong, so the power transmitted to the pendulum will be lacking. Next, check the pivot holes of the pallet arbor. Hold the pallet arbor near the pivot hole with the tweezers and lift it up and down. If the clearance is excessive make a note to bush the hole; do this for both pivots.

Now begin stripping the rear plate so that the movement can be rested on its back; this makes working with the front plate much easier. To do this you start by undoing the two screws which attach the back cock to the movement; there is no need to mark its position because this is usually set up incorrectly anyway, or is pinned. With the back cock off you can check the condition of the pallet arbor pivots and the pallet faces. You are looking for irregular shapes and deep scores. Running a fingernail over the pivots and pallets should allow you to feel any roughness, and can be a good indicator as to whether or not that mark is a groove or a shadow. Place the components in a small storage tray.

If you are working on a striking clock you will have the hammer to remove next, unless it is mounted between the plates as in most English clocks. Hammers are usually pinned to their arbor but occasionally have a cock supporting them. Unpin or screw the hammer and pull at its hub to remove it from the square. Check where the hammer joins to the hub because this joint can come loose; if this is so make a note to repair it later. Your striking clock may also have a countwheel on the rear plate. Remove the taper pin and hold the countwheel near its centre; if you cannot pull it off by hand then get a pair of hand removers underneath and prise it off its square carefully, being sure to use both hand levers so as to not bend the arbor. Do not pull from its outer edge if it is a tight fit or you may end up doming the countwheel.

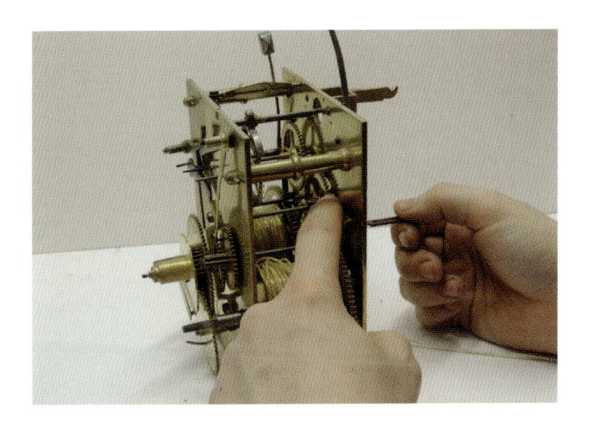

Apply power to the third wheel with a finger as you lead the escapement through by hand.

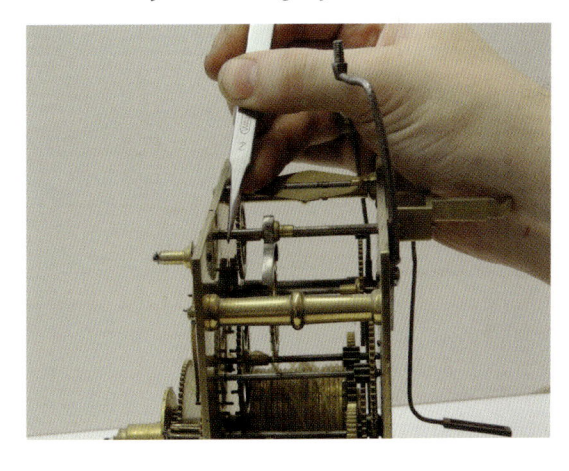

Test the pallet arbor for end-shake and side-shake before disassembling the back plate; this is easily missed.

To remove the pallet arbor, feed it through the provided hole in the back plate; this can be a fiddle but the hole is always big enough.

Remove a tight countwheel with hand removers.

Use these handy movement holders to raise the movement and hold it steady as you work on it.

Chiming clocks can have their chime barrels and hammers removed now; it is not necessary to mark the position of the barrel as it is easy enough to find during reassembly. It is a good idea to mark the hammers numerically as they come off, as they are bent to shape to reach the gong they are intended for. These assemblies usually come apart with one or two screws. The studs on which the hammers and frame rest are usually threaded into the plate. If so, I like to remove them as it makes for an easier cleaning process, but they are sometimes riveted to the back plate in which case you should leave them in place. You should always take the hammers off their post; this is so that you can clean the pivot properly and avoid sticky hammers which result in a quiet or erratic tune.

Now that you have cleared the back plate you can lay the movement on its back. Make sure that the bench is clear and that nothing will scratch the brass plate. If there are studs still in the back plate, then using movement stands will hold it a few centimetres from the bench top; alternatively, a small section of thick cardboard tube makes a nice versatile movement stand (mine came from a roll of wide masking tape).

With the clock facing up you can strip the front plate. It is best done in this position so that the parts cannot fall off and roll away. The front plate design can vary hugely, but with a basic understanding of how the components interact, their exact layout becomes less important. The order in which you strip the front plate is dependent on how it is laid out. For example, on a longcase clock the lifter is usually lifted by a pin on the front of the minute wheel, therefore it is on top and must be removed first; on a French movement the lifter is lifted by a pin on the rear of the cannon pinion, so it is the last component to be removed. If it is your first time stripping a front plate it is worth taking the time to write a list in the order in which you remove the components, and when reassembling the clock, just work backwards through the list.

Front plate components are usually held in place with a small taper pin. On some carriage clocks these pins are so small that you may need to modify a standard pin by filing to reduce them, so if they are in good condition, keep them in a small container to be put back. Alternatively, on modern clocks a C-clip is used. In this case a small groove is turned into the stud and the clip pressed on. Removing these is as simple as pressing on the open end by squeezing with the pliers or tweezers between the open end and the stud. A word of warning: small C-clips will shoot across the room when they release, so put a finger in the way to hold them in place as they unclip.

English rack-striking front plate; note the lifter position.

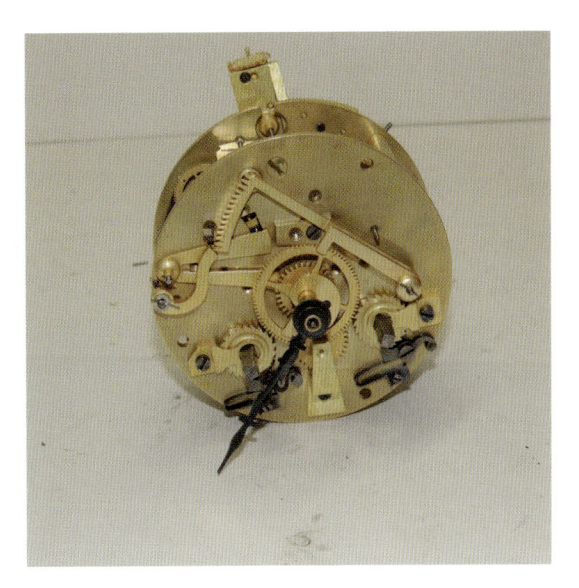

French rack-striking front plate; note the lifter position.

Gathering pallets are fitted either to a square arbor or press-fitted to a round arbor. To remove them from a square arbor, remove any pins or nuts holding them in place and pull straight out along the arbor axis. If the pallet will not come off, use steel hand levers; never lever with a screwdriver, which will bend the arbor, resulting in breakage. For gathering pallets pressed onto a round arbor I find it is best to hold the wheel steady between the plates with one hand, then the gathering pallet can then be twisted and pulled to release it.

Cannon pinions come in several designs. On English movements they will generally lift straight off, followed by the small brass leaf spring which controls the hand-setting friction. On French movements they are pressed onto the centre arbor to provide friction, which will come off when pulled straight along the axis of the arbor. Modern clock movements have a small pinion pressed on, which is an interference fit and should not be removed unless necessary.

There are a lot of springs on the front plates of modern clocks, fewer on antique clocks. I find that, provided the movement is not too dirty, it is best to leave them in place for cleaning; this saves you from having to mark their positions and readjust their tension when replacing them. If they cover any pivot holes or show signs of damage, they should be removed and inspected.

Parting the Plates

With the front and rear plates stripped, you can inspect the gear trains for pivot wear, depthing issues and excessive or insufficient shake. These points are best noted before the train is stripped. To do this, take a pair of tweezers and grip the escape arbor close to the rear plate, pull gently up and down and side to side and note any excessive movement in either direction of the rear pivot in its hole. Next, hold the escape arbor near the front plate and check the front pivot for wear in the same way. Make note of any wear as you work your way down the train, and repeat the process for the striking and chiming trains. Finally, rest a finger on the lowest wheel in the train and rotate it back and forth; as you add and remove power to the train you will easily see worn pivots jumping around in their holes. Double-check any pivots you are not confident about before moving on.

Next, check for end-shake. Grasp each arbor with your tweezers and move it back and forth; every arbor should have noticeable movement to it, but not significantly more than any other arbor respectively. There are no precise measurements to give here.

Test all arbors for end-shake and side-shake before parting the plates.

If the end-shake of one or two arbors is excessive or insufficient, check if they have previously been bushed. If so, the bush could have been set too deep or too shallow in the plate.

You are now ready to strip the plates, but first you need to figure out which plate to remove. Antique clocks, with a cross pin and riveted pillars, are simple; fit the movement stands so that the plate with the pillars riveted on faces down. You can then remove the taper pins and lift the opposite plate straight up. For most clocks of this type, this means that the back plate is to be face down, but for carriage clocks in particular, the front plate can be placed down and the rear plate unpinned.

For pillars which are bolted on, you have the choice of removing the front or rear plate. Check the centre arbor design to figure out which plate to remove. For example, you saw during front plate disassembly that modern movements have a small cannon pinion which is a press fit to the centre arbor and cannot be easily removed. In this type of movement, lifting the front plate would lift the centre arbor with it, and with the centre wheel being the closest to the back plate, the wheel would knock all the other wheels out of place, putting unnecessary stress on the pivots. Have these movements set up face down, and remove the back plate.

Whichever plate has been removed, it is now time to remove the wheels and arbors, as well as any hammers or lifting levers. The order in which you remove them is obvious; start at the highest wheel and work your way down. Modern centre arbors where the friction for hand setting is in the centre wheel should be the first to be removed; you will not be able to remove the rest of the train otherwise. To remove it, put some pressure down on the friction spring to lighten its grip on the fastener, and either press out the tape pin or remove the C-clip. The friction spring and wheel will then lift off (you may need to 'wiggle' it a little to get it past the groove in the arbor). Leave the centre arbor in place; you can remove it last, but only if the front pivot appears excessively worn.

As you lift out the components, inspect them for wear. Pinion leaves often become severely pitted in older clocks; if this is so you will need to make a repair before you can reuse the part. Pivots become scored; they should be shiny, although dullness is not a problem as long as they are smooth. Check them with a loupe and run your fingernail over the surface to feel for roughness. Arbors need to be straight, as do pivots, with the wheels perfectly square to its axis; to check this, spin the wheel between your fingers and check for wobble. Finally, check the teeth of the wheels to spot uneven divisions and bent teeth. Wear in the tooth tips is common; if it is no more than a slight flat spot, it is unlikely to cause a problem, but I have seen them worn through more than halfway.

Modern friction drive can be released by pressing down on the friction spring to relieve pressure on the pin, which can now be removed easily.

As you remove the components, it may be a good idea to label the chime components of modern clocks, as they can appear very similar to others in the train, and can be put back in the wrong position. It is wise to label all three second wheels to avoid confusion.

Labelling wheels is common, and you may find that the clock you are working on is already labelled. It is often the case on older clocks that you will see various letters scratched onto components in inconspicuous areas. Commonly used letters are: G for going train, W for working train (same as going train), S for strike train and C for chime train. You may also see dots or numerals scratched or punched into the components, usually indicating 1 for going train, 2 for strike train and 3 for chime train. If you can understand the previous marks, it is best to not cause further damage by making your own. As you do more repair work, you will need to label fewer components; I have now completely stopped labelling them, which is great from a conservation point of view.

Barrels

The power supply needs to be handled with special care, especially in a spring-powered clock.

For longcase and Vienna regulator clocks, the gut-line can be removed as you disassemble the barrel. The greatwheels of these two clocks are often held on with a keyhole piece and either a screw or pin. Before removing the screw/pin, check the fit between the greatwheel and the barrel; it should not rattle around. If it does, remove the screw/pin and slide the keypiece out of the way; the greatwheel will come loose. Check whether the looseness is in the greatwheel's centre hole by checking for side-shake on the arbor. If the side-shake is minimal, then it is just a loose keyhole, which you tighten by dishing slightly. If the greatwheel is loose on the barrel arbor (it should spin freely, but not rat-

Longcase, Vienna regulator and fusee key pieces can be released by removing the screw or pin and pushing with your fingers.

Dishing the key piece with the ball of a hammer will increase its friction and tighten a sloppy greatwheel.

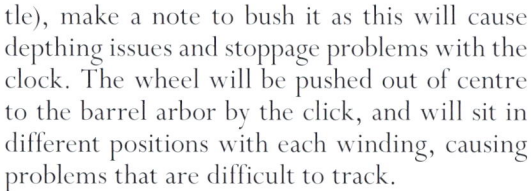

Do not forget to remove the knots from inside the barrel; this hand hole makes it a simple job.

tle), make a note to bush it as this will cause depthing issues and stoppage problems with the clock. The wheel will be pushed out of centre to the barrel arbor by the click, and will sit in different positions with each winding, causing problems that are difficult to track.

With the greatwheel off, you can remove the gut-line; Vienna barrels come to pieces to make this easy, and the process does not need explaining step by step. For longcase clocks, just cut the line and proceed to tip the barrel up to remove the knotted end of the line through the hole provided (a bit of shaking and persistence and you will clear the barrel of knots in no time). You will often find years' worth of gut-line knots in the barrel, but you should remove them because they can occasionally fall out and jam the winding mechanism.

For spring-driven clocks, the barrel will contain the spring, barrel arbor and the barrel cap which holds it all together. Barrel caps 'clip' on. Check the barrel for wear of the hole around the arbor; they wear round and can become very sloppy. Check that the barrel cap is a good fit and not loose, making notes for any repairs needed. I find the best way to remove the cap is to knock the barrel arbor straight down onto a wooden surface (a surface softer than the steel barrel arbor to avoid damaging it); this will push the cap off from the inside. Wrap the barrel in a towel to protect your hand and give it a fairly good whack – some can be a bit tight. Alternatively, you can prise the cap off with a screwdriver through the hole provided, but it is common to see unsightly chips, dents and scratches around this hole, which is the reason I suggest the other method.

Take the cap off and place it to one side, then, with a pair of pliers, remove the barrel arbor. You may need to give it a twist as you pull to unhook the spring; if the spring starts to pull out as you pull the arbor, stop and reassess the situation; it would be dangerous to yank the spring straight out. In this situation I usually pop a screwdriver blade between the arbor and spring as I twist the arbor with the pliers; this usually separates the connection. Check the barrel arbor for damage; it is common that the square is slightly rounded over, in which case make a note to repair it later.

The spring should be removed with a mainspring winder, but first, visually check it for damage. A broken spring cannot be removed with a mainspring winder so if you can see that it is broken, hold the inner edge with pliers, wrap the barrel in a rag and wear leather gloves as you pull the spring out. The rag will catch the spring and gloves will protect your hands.

If the spring is larger than 20mm wide or thicker than 0.35mm, then it is not recommended to remove it by hand. Instructions for the mainspring winder vary by design, so I will not go into detail here. To remove the spring by hand, put on a pair of thick gloves and grasp the

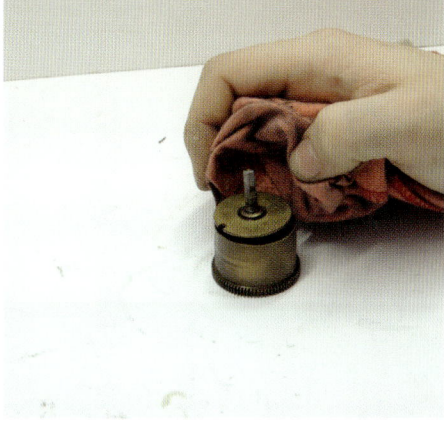

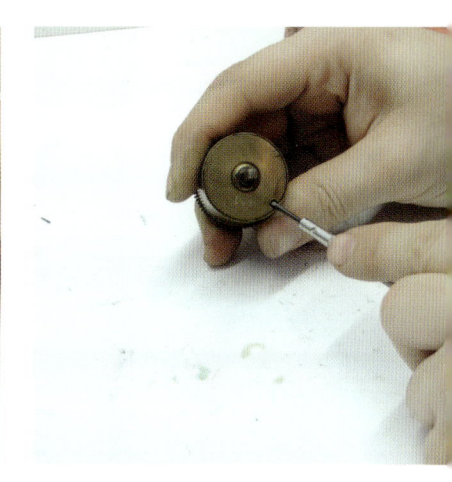

Release the barrel cap by tapping the arbor on the workbench.

The barrel cap will pop right off with no damage.

This method of releasing a barrel cap often results in a scarred and dented barrel.

barrel in a tea towel or rag. Grip the spring near its centre with a pair of pliers and pull straight out (I like to twist as if winding the spring at the same time to reduce friction – this way the spring pulls out much easier). Only pull out the first coil or two, then allow the rag to cover the spring, put the pliers down, and slowly unravel the spring within the rag. This will contain the spring if it releases at speed and the gloves will protect your fingers from the sharp edge of the spring. When the spring is unravelled, it will remain hooked to the barrel; grip the spring and push its outer end back into the barrel to unhook it.

Now you can inspect the spring for damage. A spring does not need to be broken to be replaced. A 'set' spring will have very little power or duration between winds. As you wind the clock, the spring tightens against the barrel arbor; a tall barrel arbor hook will distort and wear the spring, which will show as dents or kinks along the inside of the spring. If the spring shows signs of this it can cause the spring to 'bind' against itself or could even be a weak point for a crack to develop. If the barrel hook is excessively tall, consider reducing it in height to prevent future damage. The outer hooking eye of the spring should have a smooth continuous profile; the shape varies but is often oval or pear-shaped. If the edge appears rough or shows any sign of cracking it needs to be attended to, so make a note to repair or replace the spring.

Finally, check the barrel. Start by inspecting the teeth. If it is a going barrel type of clock (rather than a fusee) make a note of any repairs. Then check the barrel hook; it should not be loose, standing proud on the outside of the barrel or, in the case of 'punched' barrel hooks, showing signs of cracking at its base.

Fusee clocks can be treated as you would any spring-driven clock barrel, with the exception of having no teeth to inspect. The only addition to the barrel is the hole into which the chain hook or gut-line pass. This is worth checking for damage and noting; it is not usually a problem, but making new holes is an option if the originals are severely damaged. Fusee springs cannot be removed by hand safely, so do not risk it. If you do not have a mainspring winder, take the barrel to a local repairer to have the spring removed.

The fusee cone and greatwheel come apart with a screw and keyhole piece as with longcase and Vienna regulator clocks; the exception is that the click work is contained within the greatwheel and may fall out when the wheel is removed, so do this over a table or at least a large cloth on your lap. Inspect everything as before, including looseness of the greatwheel, and make notes regarding repairs.

Chapter 8

Materials and Cleaning

MATERIALS

Steel and brass are the main materials used in the manufacture of clocks. Although you do not need to be an expert in material science to understand how to repair clocks properly, it does help to have some basic understanding of the materials.

Steel

Steel is a composite of mostly iron and carbon, with carbon making up no more than around 2 per cent of its weight. Different grades of steel are available for different jobs, and in engineering, selection of the right grade can be critical. Clocks can generally be repaired using silver steel, a tool steel which can be heat-treated easily at home.

Iron, and therefore steel, is magnetic, which can affect the running of a clock. A magnetized hairspring, for example, will cause the coils to adhere, throwing timekeeping out of the window. You can buy demagnetizers to correct this; you should also regularly demagnetize your tools.

Steel can be cut with steel tools but the tool must be harder than the work piece. Generally speaking, the hardness of the steel is referred to by the colour of its temper. For example, blue steel is readily available from material suppliers for turning staffs; it is blue because it has been heat-treated.

You can harden carbon steel by heating it until it glows red, then quenching in oil or water to cool rapidly. Not all steels harden, so test a small piece first if you are unsure what type you have to hand. Hardening affects the steel on a molecular level and the result is a material

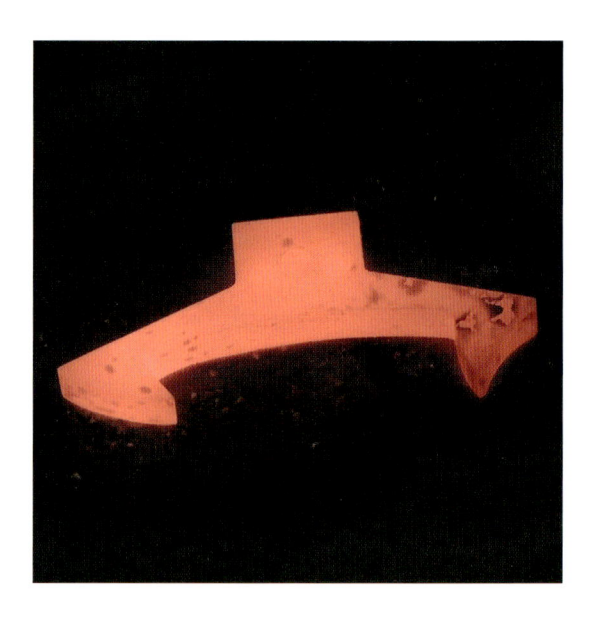

Hardening a steel recoil anchor by heating to a red glow before plunging into oil or water.

Tempered anchor showing harder straw-coloured pallets for wear resistance, but a less brittle blued anchor.

which is brittle but resistant to scratches, wear and bending.

To relieve some of the brittleness, while retaining some of the sought-after hardness, the steel needs to be tempered. Tempering steel requires reheating it to a specific temperature after it has been hardened. As you heat steel its surface changes colour, and the more you heat it, the 'softer' the material gets, so you can use its colour as a guide. For example, 'straw'-coloured steel, the first or coolest colour to appear during tempering, is harder than 'blue' steel, which is tempered at a higher temperature. The order is:

- silver (plain steel)
- straw
- brown, or dark straw
- purple
- dark blue
- light blue
- grey
- grey fading into red
- red
- orange
- glowing yellow.

Once the material reaches the point of red, and left to cool slowly, it is annealed. Annealing a material removes all of the stresses internally, and effectively returns it to its pre-hardened and tempered state.

To 'blue' a pair of hands, steel which has been polished or grained, and is clean with no oxidation or oils (including those from your fingers), can be heated to blue using an open flame or hotplate.

In clock repair, you will largely be working with soft silver steel which can be heat-treated to blue once cut, or blue steel which requires no further treatment. Very occasionally you may choose to heat steel to red for hammering, because it is very malleable in this temperature range.

Brass

Brass is an alloy of copper and zinc. It is a very malleable material and is non-ferrous (contains no iron), meaning that it is not magnetic. There are many types of brass available so selection of the right material is not easy. Engraving brass has a small amount of lead added to make it easier to cut; it is generally accepted as a suitable brass for clockmaking these days, although I am sure there is some debate on the topic. Brass is susceptible to stress corrosion cracking.

Brass is a work-hardening material, which means that to harden brass it must be hammered, drawn or rolled under high pressure. When making components such as a rack-tail or a friction spring, which must be springy, it is important to use hardened or semi-hardened brass. You can harden brass at home by hammering it all over and then refinishing the surface to remove the marks.

There is no tempering of brass, but it can be annealed by heating until it just begins to turn red; do this in a darkened corner of the workshop so that the glow is immediately obvious. Brass has a low melting temperature. When the brass is hot enough, it can be left to cool naturally or quenched; from my experience and research it does not matter which.

Brass does react poorly to the presence of natural bodily oils, and you will often see fingerprints etched deep into clock plates. To avoid this, any fingerprints should be immediately wiped from the brass with a brass polishing cloth. Wear finger cots or gloves when working with clean brass.

Brass can be cut with any steel tool, provided it is sharp.

CLEANING

Preparation

Sometimes it is best to clean a clock before commencing with the engineering work, especially if the clock is very dirty or wet with oil. If you do this, then it is not necessary to be so thorough first time around but remember to re-clean the movement after repairing and, before assembling, to remove all fingerprints and oils which will etch the brass.

Cleaning a clock movement is a multi-step process. First, roughly hand-clean the components, then ultrasonically clean them, and then

Use pegwood to scrub the roots of dirty pinions prior to cleaning for a much better job.

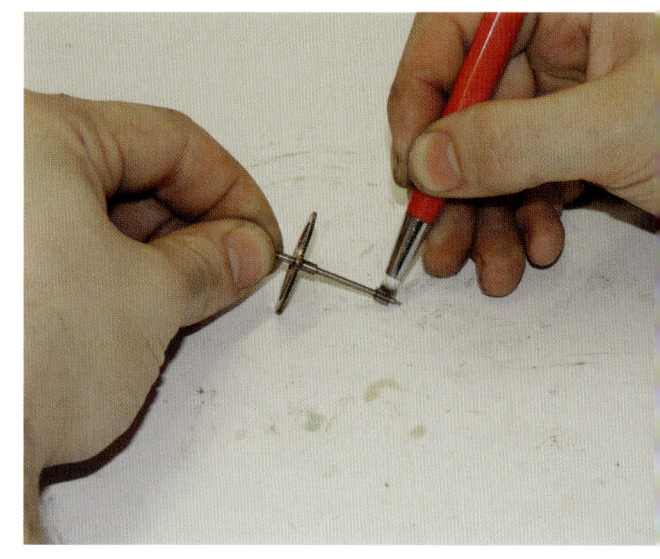

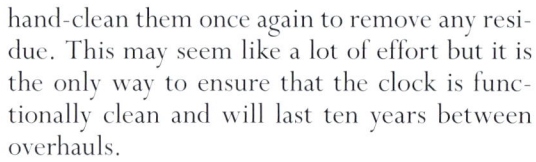

Scratch brushes can be used instead of pegwood for cleaning pinion roots and are also good for removing rust.

hand-clean them once again to remove any residue. This may seem like a lot of effort but it is the only way to ensure that the clock is functionally clean and will last ten years between overhauls.

As you disassemble the movement, you will probably notice thick deposits of dirt at the roots of the pinions, dried and thickened drips of oil on the plates, thick dried lumps of grease on the mainspring or excessive amounts of solder, or glue holding components in place. Before you can ultrasonically clean the movement, you should do all you can to reduce this by hand. Ultrasonic cleaning is excellent but has its limits, and will not clean thick dirt from the roots of pinions or remove adhesives and solder. Because of this you always start with hand-cleaning.

After you disassemble a clock movement, take a few minutes to pick at it with a piece of pegwood. Sharpen the pegwood to an appropriately sized point and scrub any dirt from the roots of the pinions and the sharp corners of the wheel crossings and teeth, and wipe off any dry oil from the sliding surfaces of levers and pins. If you can see any significant staining on the plates, try to wipe it off, using a little brass polish if needed. If it will not budge, use a fibreglass scratch brush (although be prepared to re-

polish after doing this). I would suggest wearing latex gloves and working over a bin or tray when using scratch brushes, as the fibreglass bristles break off during normal use and find their way into your skin.

With the thick deposits removed, you need to assess whether or not the plates have been lacquered. Ultrasonic cleaning will make a mess of lacquered plates, half-stripping the lacquer and turning the remains a milky white. If you have your suspicions, take a graver and scrape the edge of the plate; the lacquer will flake off and leave a clean patch of brass. If the plate is lacquered you need to make a decision dependent on its condition. If the lacquer is brown and dirty or chipped and unsightly, then it will have to be removed. Alternatively, if the lacquer is clear and in good condition, giving the appearance of clean and shiny brass, then it should be left alone. Some clocks, especially common quarter-chiming movements made by Elliott, have a 'cracked' lacquered appearance. If this is intact then it should be left alone. If it is scratched and unsightly, or appears damaged by previous cleaning, then it should be removed and the plate lightly grained.

A word of caution; when removing the original lacquer, such as in the Elliott movement previously mentioned, ensure that there is no writing printed onto the components, as found on the chime selector knob of these movements. If there is, carefully clean these components by hand.

To remove the lacquer, I favour soaking the plates in a strong water-based cleaning fluid, periodically removing the plates to scrub them with a stiff brush. Wear thick rubber gloves while you scrub and do it in a well-ventilated area. Slowly the lacquer will loosen and fall away from the plates, after which you can rinse the plates off with soap and water and resume the usual cleaning process.

There have been occasions where the lacquer would not come off, and I have had to resort to paintstripper. It has never had an adverse affect on the brass, but do ensure that it has been thoroughly removed when you have finished.

If you decide not to remove the lacquer, then you will need to thoroughly clean and degrease the plates by hand. For this I use washing-up liquid. Use a good amount and scrub well; when you are done make sure that you rinse and dry the plates properly; a hairdryer is preferred after towelling off.

Ultrasonic Cleaning

It has become common in clock repair work to use ultrasonic cleaning tanks to thoroughly clean the movement. Ultrasonic tanks are expensive and not necessary if you only intend to repair the occasional clock. However, if you are planning on clock repair becoming a commercial activity then a significant amount of time can be saved by taking this approach.

Ultrasonic cleaning uses high-frequency sound waves to scrub tough contaminants, oil and grease from any intricately shaped component. These sound waves are at the top of, or above, the frequency range of human hearing and cause the cleaning solution to vibrate with such intensity that microscopic bubbles are formed (known as cavitations). The bubbles contain a proportionally large amount of energy which is released as a jet of liquid in a tiny implosion as the growing bubble reaches its critical mass and collapses. The tiny jet travels at high speed and blasts contaminants away from the components you are cleaning.

A tank of around 5 litres (1 gallon) should cover most of your clockmaking needs, although I still occasionally find that ours is too small, and I am forced to clean large movements in several small batches, which means decreasing efficiency and increasing workload. If you intend only to work on small French movements and carriage clocks, then 5 litres may be excessive. You should determine your own needs and talk to the supplier to find the right tank.

Ultrasonic tanks regularly show up at horological auctions (held regularly by BHI branches throughout the UK), and this is where my tanks are sourced. Buying at auction is a risk because you cannot be sure of their condition, but they are repairable if you are reasonably competent with electronics. My latest crisis involved changing a burnt-out rectifier for less than £3.

When cleaning ultrasonically, you use three separate fluids: one ammoniated cleaning fluid (alternatively you can choose an ammonia-free fluid) and two alcohol-based rinsing fluids. The first rinse removes the majority of residue, and the final rinse ensures absolute cleanliness.

You need a way of getting these fluids into, and back out of, the ultrasonic tank every time you change from cleaning to rinsing. Most tanks come with a drain plug for this purpose; with the components in the basket, you can drain and refill the tank easily, but slowly. We use a second, plastic container, which we place within the ultrasonic tank. The ultrasonic tank is filled with water to the indicated level, and when we fill the second container with cleaning fluid, the ultrasonic vibrations are passed through the water and into the container of cleaning fluid. The clock components are placed directly within the plastic container of cleaning fluid for the duration of the cleaning cycle, and upon completion we lift out the inner container, and pour its contents through a sieve and funnel back into the bottle, and leave the components in the container.

The plastic container is then placed back into the water-filled ultrasonic tank, with the movement and components still inside, and topped up with our first rinse solution; we run the machine for five minutes or so in this rinse. After draining off the remains of our first rinse back into its container, we repeat the process with the second, much cleaner rinse solution.

Components in the ultrasonic tank ready for cleaning.

Remember to fully submerge all components in the cleaning fluid or a tidal mark will stain them.

The amount of time to leave the components submerged is dependent on several factors: the original condition of the components, the age/cleanliness of the cleaning solution, the type of cleaning solution and the temperature of the solution, etc. However, the following can be used as a guideline;

- cleaning solution – ten to twenty minutes
- first rinse – five minutes
- second rinse – five minutes.

If the plates are particularly dirty, I find it helps to increase the cleaning and first rinse periods, and to repeat the process if the results are not satisfactory. It is wise to lightly scrub the plates with a stiff bristle brush halfway through. This removes the top layer of dirt which has been softened so that the ultrasonic waves and detergents can begin to clean the next layer down.

When cleaning a movement in this way it is important to submerge the plates fully; if a section is left protruding then it will not be brightened by the ammonia and the result will be a tidal mark showing clearly where the cleaning fluid ended.

Some ultrasonic cleaning tanks come with a temperature dial to warm the fluid. Warming the detergent solution does help its cleaning properties, but it is not necessary to do this on every job. The extra heat speeds up evaporation and when heat is used with the rinsing solution, the plates can become discoloured. Heat the dirtiest movements during the initial clean to 50 °C or so (a temperature I have found to work excellently without being unnecessarily high) and leave the others running cool. The

Drying on eggtrays with a hairdryer works wonderfully; just to be sure I then place them in front of a fan heater.

ultrasonic process does produce some heat of its own, and if left for long enough the solution will get hot. Remember that the components may be uncomfortably hot to handle when removing them from the tank.

After the final rinse has been drained back into the bottle, empty the components onto an absorbent material (eggtrays are very useful) and blow dry. The final rinse evaporates and within a few minutes the components will be completely dry. Avoid touching the brass with your skin once is has been cleaned; use a tea towel or clean rag and place them carefully back into the container to be transported back to the workbench.

Hand Cleaning

If you have chosen not to buy an ultrasonic cleaning tank, then you can do just as good a job by hand cleaning. Buy some water-based cleaning solution and make up about 2 litres (½ gallon) as per the instructions on the bottle.

After picking away the thickest dirt deposits by hand, the purpose of this second stage is to degrease the movement and brighten the brass. Place the components in a bucket or spare washing-up bowl and pour in the cleaning solution. Leave it to soak for ten to twenty minutes; this will allow the dirt and grease to soften, making hand cleaning significantly easier. After ten to twenty minutes, put on a pair of thick rubber gloves and pull the components out one by one. For each component, take a scrubbing brush with firm bristles (but not a wire-bristled brush) and scrub away any grime. Use a toothbrush to get into the pinions and make sure that everything is clean. You may find that you need to leave parts soaking longer to soften the dirt, but eventually you will be left with spotless components.

Pour the cleaning fluid through a sieve and funnel, back into its container, keeping the clock components in the bucket. You can keep the ready-mixed cleaning solution for later use; it will last several cleans before it needs replenishing. Top up the bucket with warm water and a little washing-up liquid and clean the components once again, but this time with a soft sponge. Pour away the washing-up water, being careful not to lose any parts down the drain.

I find the best way to dry the components is to boil the kettle and fill the bucket to immerse the components in hot water, leaving them to absorb some of the heat from the water for a few minutes. Pour away the hot water and empty the components onto an absorptive material such as a large eggtray or pile of rags. Towel-dry the components thoroughly and quickly. You will notice that the heat from the boiled water will evaporate the remaining water leaving perfectly dry components. Just to be certain, I leave them in front of a fan heater for a further ten to twenty minutes, or use a hairdryer for the same effect.

Chalk-Brushing

Back at the workbench you can begin chalk-brushing the components and pegging out the pivot holes. The chalk brush is just as it sounds; a medium-hard, hand-held bristle brush is drawn across the surface of a piece of chalk. The actual role of the chalk seems to be of some debate, but to me it serves several roles: it cleans the brush, it absorbs any liquids which may be present on the components, and it acts as a mild abrasive.

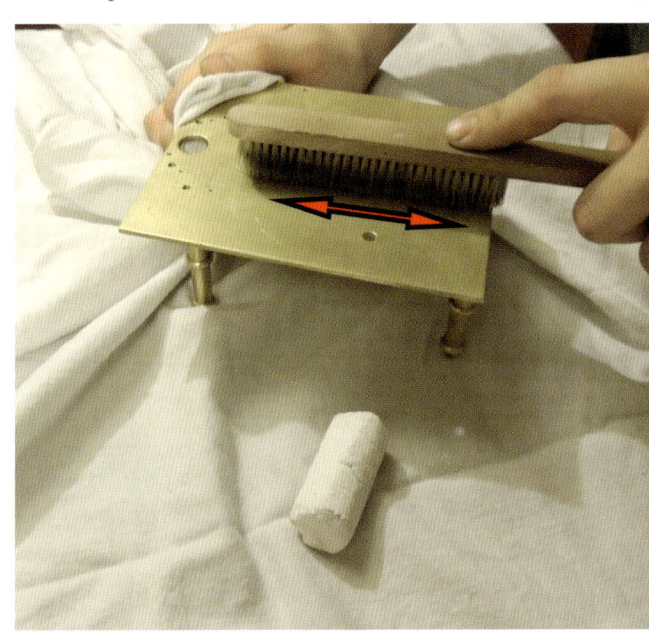

The chalk brush is a traditional cleaning method which brightens the plates and scrubs away any remaining residue.

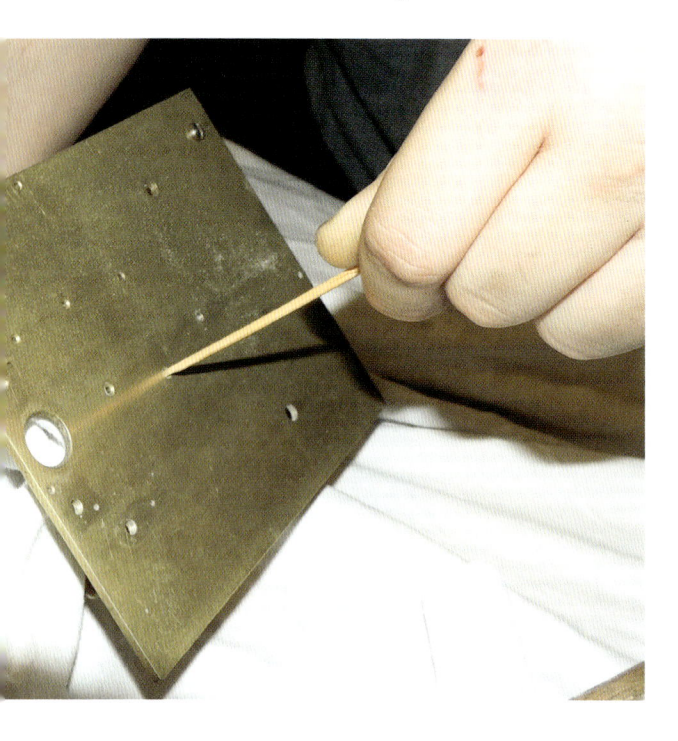

Always remember to pegwood the pivot holes until they are spotless.

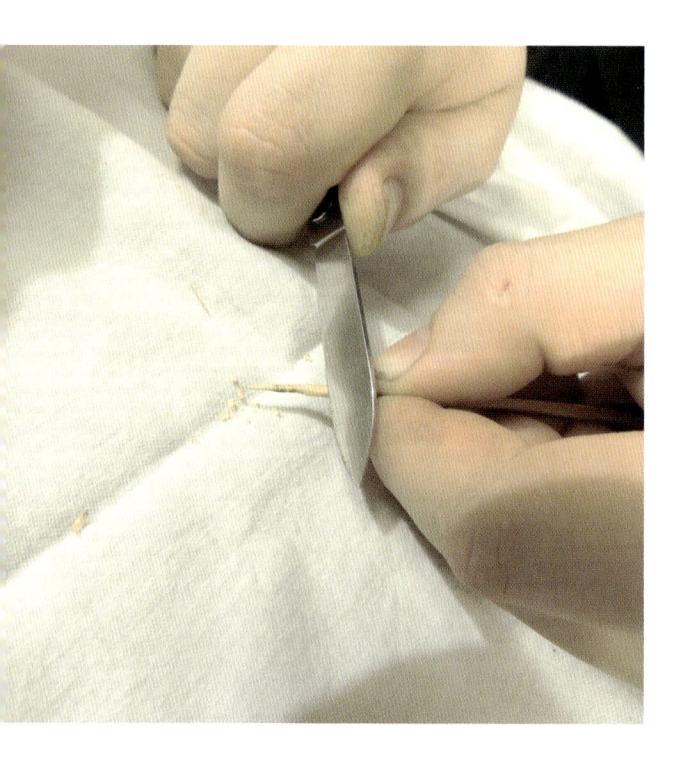

Keep repeating this sharpening process until every pivot hole is spotlessly clean.

The charged brush is used to thoroughly scrub each component until all traces of contaminant are removed and the brass is left shining. As you scrub the components, hold them in a clean cloth to avoid touching them with your skin, or wear gloves if you prefer.

After all the components have been scrubbed clean with the chalk brush, you need to clean out any residue from the pivot holes of the movement. Any dirt left in the pivot holes will mix with the oil and increase the speed of pivot wear. Take a piece of pegwood and sharpen it to a long thin point. Insert and twist it inside each pivot hole, withdraw the pegwood, re-sharpen it and repeat the process until it comes out clean. Turn the plate around and repeat the process from the opposite side. Repeat this process with every pivot hole.

For the larger pivot holes you may not have any pegwood large enough to clean them. I like to take a strip of chamois leather and draw it back and forth through the pivot hole to clean it.

With the components chalk-brushed and the pivot holes pegged and clean, the only remaining residue is likely to be in the oil sinks. Sharpen a large piece of pegwood, but this time into a flat blade similar to a screwdriver. Round off the end of the 'blade' to match the diameter of the oil sink, place it in the oil sink and twist it back and forth.

Small components should be cleaned in the watch cleaning machine, using the same three-stage process as with the ultrasonic cleaning tank. If a watch cleaning machine is unavailable, then the components of platform escapements can be cleaned in a small jar of lighter fluid or hairspring degreaser and gently scrubbed with a soft artist's paintbrush. They can then be dried on blotting paper or using a commercial drying medium.

Polishing

Polishing components is not something I recommend too often. It involves the removal of material and eventually leads to uneven surfaces with rounded corners, and overall a reduced serviceable life. From a conservation point of view, chalk-brushing is the final operation and anything further than this is too much.

There are occasions where polishing a component or two is necessary. For example, we have sometimes had to scrub away an etched fingerprint on a carriage clock; this is common as people set the hands with their fingers rather than the clock's key. Their fingers smear acidic oils on the plates, and the process is repeated once a week for ten years or more. Carriage clocks are largely visual, with a glass door at the rear intended for viewing and enjoying the intricacies of the movement, so it should be spotless.

Polishing is a simple process when taken step by step, and a mirror finish can be easily accomplished with some patience. To polish metal you often have to start by making things worse, and in the example above that is exactly what we do. We start with a 600 grit abrasive paper, working in one direction only, concentrating on the problem areas, but lightly treating the entire surface of the back plate of the clock, until the blemishes are removed. It is important to use some common sense here; scrubbing wildly at the entire plate until one small area has had its blemishes removed may result in engravings being scrubbed away, so concentrate on the problem areas and lightly treat the entire plate only to blend the finishes into one. If a blemish is so deep that removing it involves the removal of significant amounts of material, stop. Improve the situation but do not scrub until there is an equally unsightly pit in the finished component.

After removing all blemishes with the 600 grit paper, move on to a finer 800 grit paper. The higher the number, the finer the finish. Work with the 800 grit until all of the scratches produced by the 600 grit paper have been removed. Some recommend working at 90 degrees to the scratch marks from the previous paper to make it easier to see when to stop. Then move on to 1000, 1500, 2000 and finally, 3000 grit.

After using the 3000 grit, you should have a fine matt finish. From this point you are ready to move on to finishing with abrasive pastes. It is down to your personal taste how far you choose to go here; at this stage you are removing tiny amounts of material. If you want a perfect mirror finish then you will have to work through the various pastes on the market; however, I suggest going no further than a good scrub with some brass polish on a soft cloth at this point, producing a finish which shines beautifully and will always impress.

After polishing it is extremely important to remove every trace of polishing compound and grit. I suggest that you clean the parts thoroughly in lukewarm water and washing-up liquid (or ultrasonically), then dry them thoroughly and repeat the steps of chalk-brushing and pegging out.

Polishing of turned pieces can be done in the lathe, speeding up the process greatly.

Chapter 9

Repairs

THERE IS NO 'ONE SIZE FITS ALL'

When it comes to clock repairs, there is no definitive right way, but there are certainly some wrong ways to do things. You really need to understand the theory in order to make the best choice for a given situation. So in this following section I am not showing you the way things should be done, but the methods I use for that repair. You will need to adapt the methods to suit your needs and tooling.

If you are repairing a rack-tail, you will need to understand the relationship between the length of the rack-tail and its position, relative to the rack. You will also need to have an understanding of the materials used in its construction and how to work them. If in doubt, re-read the theory on that component, think twice before you act, and never introduce a new problem to avoid correcting the original one. There is nothing worse than having to repair a striking mechanism that has had the snail filed half to death because the previous repairer did not know how to adjust a rack-tail.

That being said, I suggest you approach every repair with the following three rules in mind:

- think twice
- never introduce a new problem
- respect the originality of the clock.

WEAR GROOVES

Toolbox 1

As you disassemble the clock, check all components in frictional contact for wear grooves on their surfaces. In many cases the repair is simply to file away the wear, removing as little material as possible, and then polish the surface to reduce friction. This is common in the lifting levers of French clocks.

In components which need careful adjustment such as pallet faces or countwheel detents, the material removed by wear will need to be replaced. Micro-welding is excellent for this but is unavailable to most of us.

To replace the material you will need to reduce the surface and soft-solder a piece of adequate material in its place. This is covered later with the example of recoil anchor pallet faces, which is such a common example but which can be modified to suit most situations you will encounter.

BUSHING

Toolbox 2

Bushing is a step-by-step process which is best described by the pictures. Although not all situations are the same, you can always approach bushing a hole with these steps, modifying slightly as appropriate.

Large bushes should be riveted to the plates, while bushes fitted to a spring barrel should be soldered. Never use solder on the plates of any clock. Small bushes can be held by friction alone as they are under such small loads.

Bushing by soldering is covered later during the repair of a worn pulley wheel centre so will be ignored here to save repetition.

To accurately bush a pivot hole you first need to find the original centre for the pivot, or else depthing problems will result. To do this, place a smoothing broach into the worn pivot hole until it will go no further, hold it up to the light and you will notice a slither of light to one side of the broach; this is the result of wear. Use a needle file to remove material equal to the amount of wear on the opposite side of the pivot hole. If a needle file will not enter the hole, use a cutting broach and sideward pressure to the same effect. The pivot hole can now be bushed.

If the pivot hole has previously been bushed, remove this bush and replace it with your own, skipping several steps in the process and saving time.

When the new bush has been fitted, broach it open to accept the pivot as a snug fit; finish with the smoothing broach and a drop of oil to burnish the surface and bring the hole to size. The wheel should be free to rotate when spun between the plates.

A very torn pivot hole.

Use a smoothing broach to see in which direction the pivot has worn.

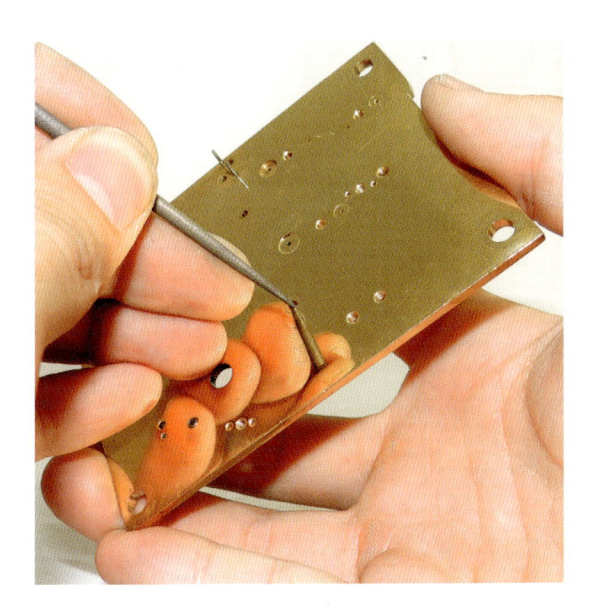

Use a needle file to lead the centre of the hole back to where it should be.

Checking once again with the smoothing broach should reveal even wear on each side before continuing.

Broach the hole at perfect 90-degree angles to the plate in all directions.

Stop broaching when the intended bush fits about halfway into the hole.

Use a flat-faced punch to drive the bush in until it is flush with the plate.

It is preferable to drill away the majority of the bush now to reduce filing and begin to create the oil sink.

Use a fine needle file to remove the remainder of the bush until it is flush with the plate.

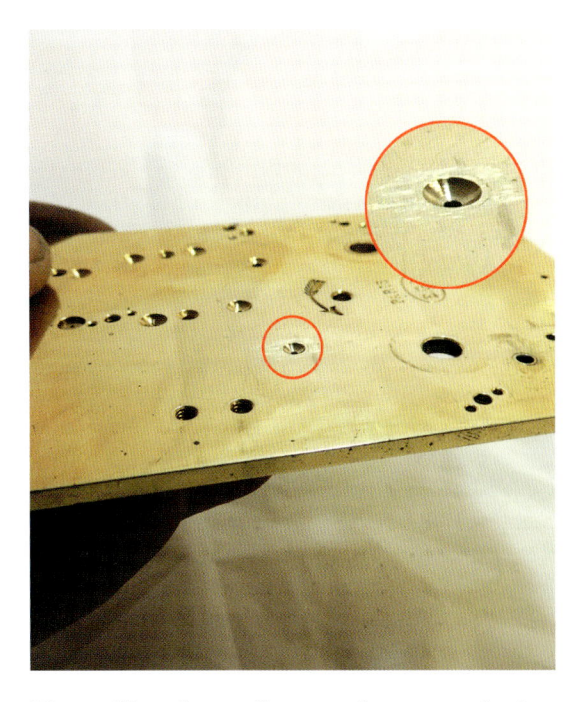

There will be only a small amount of scarring on the plate if you are careful; this will polish out without a problem.

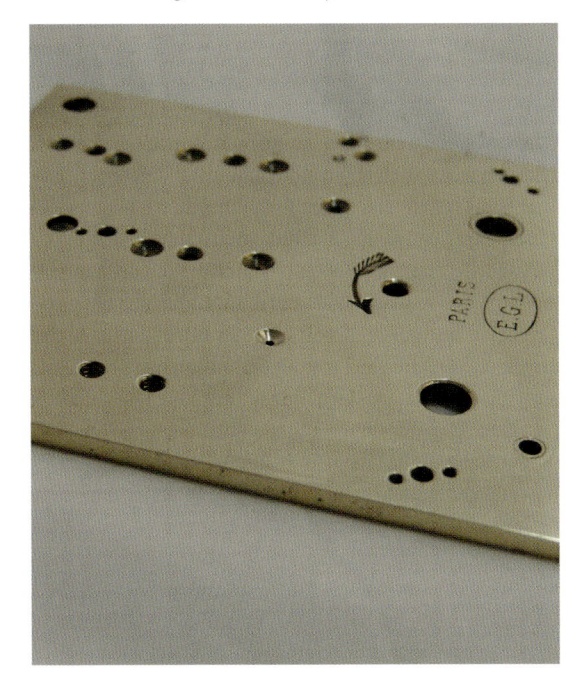

The finished bush, once broached to size, should not be obvious.

The result of a well-fitted bush should be a plate that looks as good as new.

END-SHAKE

Toolbox 2

End-shake is important to consider when repairing a clock. If you have bushed a hole then it is easy enough to adjust by knocking the bush in or out a bit, but if the pivot hole has not been bushed and the end-shake is too much or too little, then it is likely that the plates have been dished or become distorted.

If the end-shake is too much, then it is possible for the wheels to clash; if it is too little, then the arbors will bind. You should always check the end-shakes before disassembling the movement so that you can act on it, and do the same during reassembly to check that you have not left anything too tight or loose after bushing.

If the end-shake is too much, check each plate against a straightedge; if either plate bows then you will need to straighten it to put things right. To straighten a bowed plate, support it at each end on a pair of wooden blocks with the bow arching upwards like a bridge. Use a third wooden block in the centre to apply downward pressure. This is best done by leaning your body weight on the wooden block, which will allow you to increase the pressure slowly until the brass begins to give way. Make small adjustments until the bow is gone.

If the brass plate is full of stress cracks, then there is a risk of these cracks propagating during straightening, and it is best to leave them as you find them.

If you cannot straighten the bowed plate for any reason then the end-shake will have to be adjusted using another method. The best way is to fit a tall bush to the pivot hole, which will take up some of the slack. In doing this you can choose whether to reduce the end-shake from the front or rear of the movement, which allows you to push the wheel into the best position for optimum clearance, or to avoid pinion wear. As the process of bushing has already been covered I will not provide any further detail here.

If there is no end-shake and the arbor is binding, it is either because of a bowed plate (which has already been covered), a bush which you have fitted but which needs to be fitted a little deeper, or a burr on the pivot hole. In the case of a binding arbor which has not been bushed, the fix is a quick one. Take a twist drill and lightly chamfer the inside edge of the pivot hole, then lightly broach the hole to remove any burrs. It does not take a lot to remove enough material to free up the arbor, and any marks made will be removed when the pivot hole is next bushed.

REFINISHING PIVOTS

Toolbox 2

As the pivots of a clock run in dirty, dry oil, they wear away and can in some cases become significantly bottlenecked. If the wear is nothing more than light to medium scoring then you can refinish them with a pivot file and burnisher.

To refinish a worn pivot without a lathe, mount the arbor in a pin chuck, file a small notch into a block of wood held in the vice, and rest the pivot to be filed in this notch. The notch will stop the pivot from wandering as you spin the pin chuck back and forth in your fingers. Using the file end of a pivot file and burnisher (ensuring it is the right way up so the sharp corner rests against the shoulder), file against the direction of rotation to remove material from the pivot. It is very important that you keep the file square to the arbor to produce a parallel pivot, and to keep it tight into the shoulder to ensure no material is left which may bind in the hole. Take a few light cuts and check the progress; if the pivot is starting to taper, adjust the angle of the file and try again. Continue until all score marks have been removed and the pivot is parallel.

Now that the pivot is reduced and the scoring has been filed away, you are left with a rough pivot which would run with more friction than necessary in the pivot hole. Using the burnisher and a drop of oil (to avoid tearing the surface of the pivot), repeat the process in the same manner as when filing; you should soon see a mirror finish on the pivot. Burnishing both polishes and hardens the top surface of the pivot by compressing and creating a protective shell around it.

If the scoring is deeper it is easiest to do the work in the lathe. Mount the arbor in a collet and check that the pivot is running true. Grip the arbor as close to the pivot as possible, although with wheels and pinions blocking the way you will probably have to grip the arbor at one end while working on the other. In this situation you will need either a runner or a lathe steady to support the arbor nearer the pivot. These can be made or purchased. For this example I will assume you are finishing a well-supported pivot such as the rear pivot of a longcase centre arbor, holding the pinion in a collet.

Run the lathe at a low to medium speed. Approach the pivot from beneath with the pivot file so that you can observe its top surface. This allows you to notice instantly if material is being removed more from one end, suggesting that the pivot will end up tapered. Work the pivot file back and forth at moderate speed, and remember to keep an eye on the corner where the pivot and shoulder meet. When the pivot is sufficiently reduced, use a drop of oil (a thin smear should do it) on the burnisher to finish the surface. Work the burnisher back and forth faster than with the file, and with slightly greater pressure. You will see the surface brighten as it is burnished. If the surface does not take a polished finish, your burnisher needs sharpening.

Sometimes it is better to leave one or two grooves in the pivot than to reduce the diameter to such an extent that it will be left weak. File and burnish the surface to remove as much wear as reasonable; the remaining marks will tend not to cause any problems. However, a severely reduced pivot diameter will be greatly reduced in strength and potentially damaging.

When a large diameter pivot is to be reduced significantly, or a lump in the corner of the shoulder needs to be removed, it is best done with a graver in the lathe, to be finished with the pivot file and burnisher.

LEFT: *Worn pivot holes go hand in hand with worn pivots like this one.*

RIGHT: *Pivots can be filed and burnished on the edge of a wooden block; I use a bench peg for mild cases of pivot wear.*

In severe cases it may be quicker to turn away the surface of the pivot in the lathe.

Filing pivots in the lathe allows for much quicker removal of material.

Burnishing in the lathe makes it a lot easier to ensure that the pivot remains parallel.

BENT PIVOTS

Toolbox 3

Bent pivots are either the result of a dropped clock, mainspring damage or, most likely, a clumsy repairer.

When examining a bent pivot it is tempting to grab it with a pair of pliers and bend it back to position, but doing so would, in most cases, break the pivot off leaving you with a much bigger job in hand.

Look closely through high magnification at the outside edge of the bend, where the material has been stretched, and look for any tearing or splitting on the pivot. If the arbor is showing these signs of damage, it will be left weak if straightened and will ultimately break in the near future. If the pivot is damaged, it will need to be replaced. If, however, there is no clear damage or cracking then you may be able to save it.

Before trying to straighten the pivot you will need to let down the temper to reduce the likelihood of breakage. Use a blowtorch or lamp (a lighter or candle will leave sooty deposits). Play the arbor in the flame until the colour of the steel begins to change, keeping it moving to avoid the finer pivot turning red. Allow it and the shoulder to change from a straw colour to purple, dark blue, and then light blue. At this point the temper is sufficient that you should be able to safely straighten the pivot, without having reduced the temper completely.

Find a collet which properly fits the pivot and set it up in the lathe. Using a centre in the tailstock you can rotate the lathe by hand to see a magnified example of the angle at which the pivot is bent. Rotate the headstock until the arbor is pointing upwards at its highest point and gently apply downward finger pressure to the arbor until it is pointing directly at the tailstock centre. Rotate the headstock by hand and repeat the procedure as many times as needed until the arbor runs true.

Bent pivots are not uncommon; they are caused by spring damage or mishandling.

Always temper the pivot before attempting to straighten it, otherwise it will certainly break off.

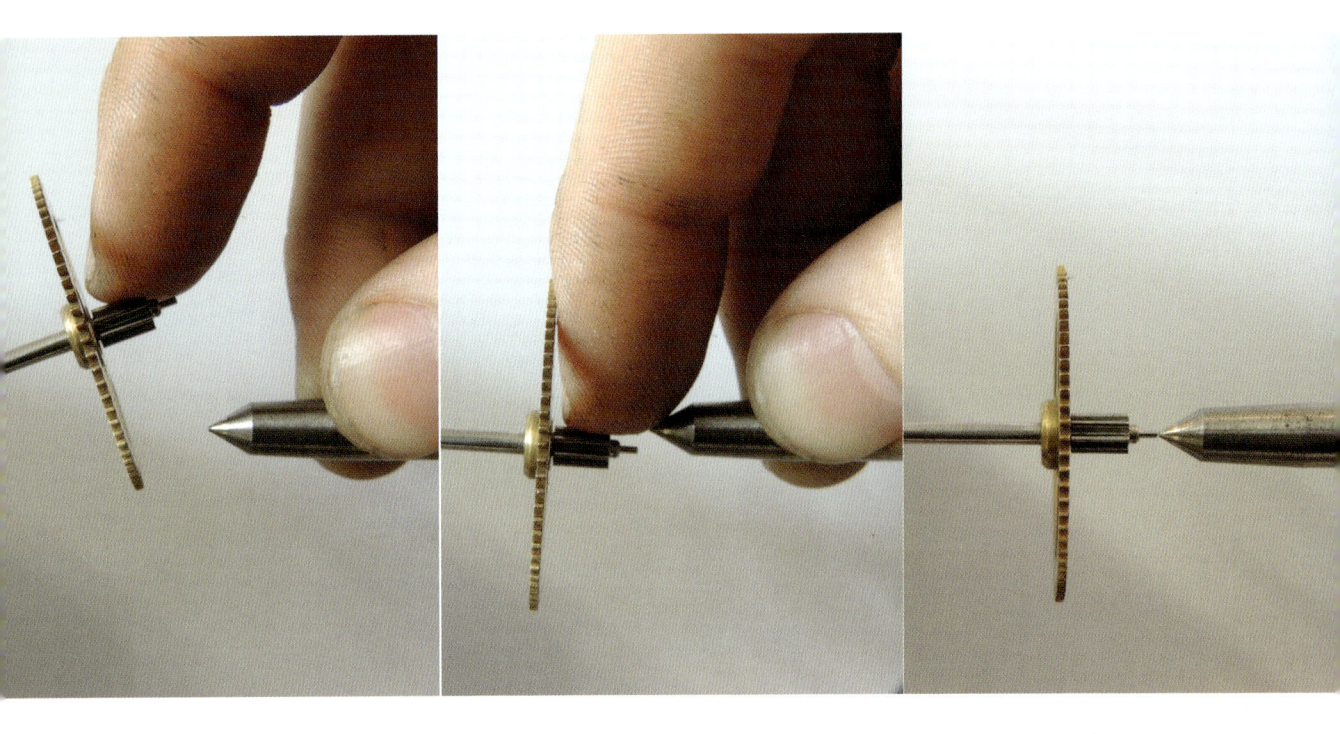

Mount the bent pivot in a collet in the lathe; this will accentuate the angle through which the pivot is bent, making it easier to spot mild cases.

Bend the pivot beyond where you need it to stay; metal has a certain amount of elasticity which you need to overcome.

The result should be a pivot which runs dead centre as you rotate the headstock.

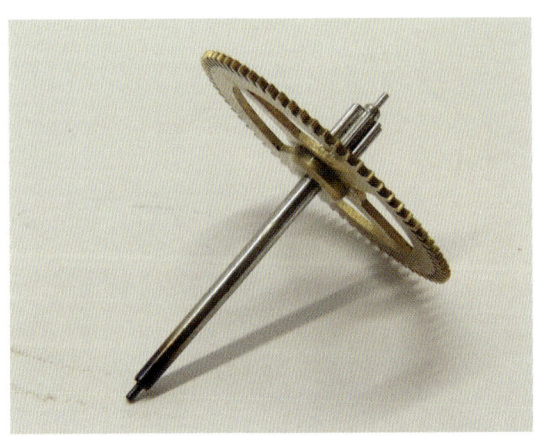

Finished pivot is now straight but may require burnishing and cleaning to remove the colour.

When polished, the pivot will look as good as new.

When this is done and the pivot is true, it will be necessary to burnish and refinish the pivot to remove any marks. It would be a good idea to harden and temper the pivot to dark blue, but as you tempered it only one stage before this, it is not entirely necessary.

Modern arbors made with free-machining steel often do not need to be tempered as the steel is left soft, but the pivot surfaces should be work-hardened by burnishing.

RE-PIVOTING

Toolbox 3

Worn pivots can become so severely worn that they can no longer be refinished by burnishing and must be broken off and a new pivot fitted. Provided this is done well it is a perfectly acceptable and unnoticeable repair, but in the wrong hands it is common to end up with an eccentric or poorly fitting pivot. Because of this, re-pivoting has acquired a bad reputation and some prefer to replace the arbor entirely.

Re-pivoting is an important skill for the repairer to learn; it will help you stand out above the crowd by giving you the ability to correct other people's mistakes, and as a beginner it will help you put right that first pivot you will inevitably break. You will need a few specialist tools including a lathe. I use the 8mm watchmaker's lathe with a specially adapted tailstock, which can be easily made with a hacksaw and drill-press. Various attachments can be bought for the lathe which make re-pivoting easier, but as a beginner, for now, I will assume you do not have these.

The first step is to test the hardness of the arbor. You will often find that older clocks have extremely hard arbors which will ruin tools and create excess heat when drilled, while newer arbors are made from a free-machining alloy of steel that can be drilled easily. To test the temper, gently file what is left of the pivot; if it is soft then the file will cut, while a hardened arbor will cause the file to slide over its surface as if you are attempting to file glass. If there is no pivot left to test the hardness on, try the file on an inconspicuous area near the shoulder.

If the pivot is hard, or if you are unsure of the hardness, then it is wise to temper the arbor and soften the steel to make it more pliable. To do this, clean the end of the arbor near the pivot to be drilled and apply heat to this end with a blowtorch. As it heats up, move it around in the flame to chase the colour blue along the arbor by at least two or three times the length of the pivot to be fitted; there is no need to soften the entire arbor. Withdraw the arbor from the

Severely worn pivots like this one cannot be salvaged and must be replaced.

Drill a hole through a piece of steel plate to make the starting of a runner.

flame and leave it to cool. The blue colour can be cleaned off with emery cloth, or with a quick soak in a pickling solution, a few seconds in a rust remover, followed by a thorough rinse in clean water.

The final preparation to the arbor is to file, stone or turn away what is left of the pivot. My preferred method is to use the lathe. Set the arbor up in a collet and, with the arbor revolving, use a diamond lap to grind away what is left of the pivot. If the arbor is sufficiently supported and there is a large amount of pivot to remove, then it may be easier to turn it away with a sharp graver.

Next you need to set up the tooling, and first you must make a drilling guide to suit your chosen drill. The drill guide we use is a tailstock, modified in such a way that it can carry a piece of steel stock and slide freely, but without play, along the lathe bed, towards or away from the headstock. The piece of steel is drilled for each re-pivoting job to ensure absolute alignment of the guide hole and to make sure that there is no play between the guide hole and the drill bit. The modified tailstock is a spare tailstock which has been cut short and drilled to receive a nut and bolt. The nut and bolt then clamp the steel stock vertically. I chose to use a tailstock because it is pre-made to slide accurately along the lathe bed. You could equally use the cross slide to clamp the steel guide piece; however, you would then loose the 'feel' which you get by applying pressure by hand during drilling.

With a piece of steel stock clamped into your modified tailstock you can select your drill bit. The size of the drill bit is dependent on a number of factors and there are no solid rules to follow. Where possible I prefer to use a drill which is slightly larger than the finished pivot; this helps achieve a strong connection between the arbor and pivot by increasing the contacting surface area as the new pivot is pressed in, and it provides enough steel to be able to finish the pivot to a high standard. However, the drilled arbor must remain strong enough that it will not split when the new pivot is pressed in, so there is a limit to the size of drill you can use. Unfortunately, some clocks were made with the ratio between pivot and arbor diameters very small, meaning that in order to drill a hole large enough to match the original pivot diameter, a very thin wall of material will remain around the hole which would easily split when a pivot is fitted. In this case it is best to use a drill bit the same diameter as the final pivot or, in severe cases, to slightly reduce the diameter of

Chamfer the edge of the drilled hole to guide the arbor to run true to it.

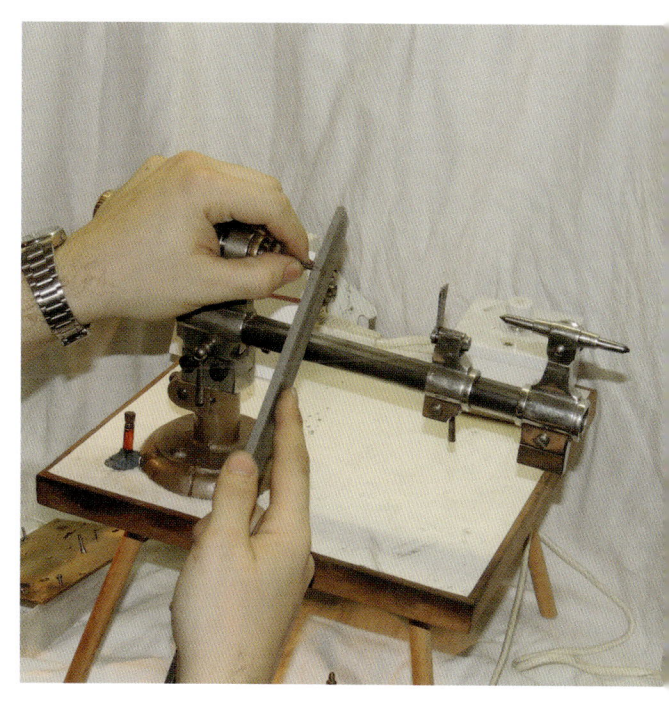

Mount the arbor in the lathe and stone, file or turn away what remains of the old pivot.

the pivot to increase the overall strength of the finished job. If the pivot hole is unworn, I tend to select a drill bit one size above that which just enters the pivot hole.

Fit the drill bit into a collet in the headstock; if it is a particularly long drill make a point of mounting it so that no more length than necessary protrudes from the collet. Start the lathe at moderate speed and slide the modified tailstock towards the rotating drill. Using light finger pressure, allow the drill to pass completely through the steel.

Now select another drill bit, this time large enough that the shoulder of the arbor will sit comfortably within the starting cone which will result from drilling a shallow hole. The size of the drill used is not critical; a drill slightly larger than the diameter of the arbor should be sufficient.

Mount the drill bit in a collet in the headstock and start the lathe at moderate speed. This time, as you slide the modified tailstock towards the rotating drill, you are aiming only to drill deep enough that a chamfer is made concentric to the through hole drilled previously.

This chamfer will act as a funnel for the arbor, guiding the shoulder to rotate perfectly concentric to the hole. Drill deep enough that the chamfer accepts the shoulder, but not so deep that the original hole is compromised. The process of making the guide piece takes no more than a couple of minutes.

Mount the arbor in a collet in the headstock of the lathe. Slide the modified tailstock up to the arbor and guide the shoulder into the cone, spin the lathe to ensure the shoulder is well seated and clamp the tailstock in place. Put a drop of oil where the shoulder contacts the guide piece to reduce friction. Finally, fit another tailstock to the lathe and mount a dead centre.

Take the original drill bit used to drill the through hole in the guide piece and put it in a pin chuck. Dip the end in cutting fluid, which helps cool the cutting action and lubricates the flutes of the drill bit. Guide the end of the drill bit into the through hole, and guide the dead centre of your second tailstock into the rear or the pin chuck. The drill is now mounted in a manner in which it will drill dead centre on the arbor and perfectly parallel with the centre line.

Set the arbor into the runner; it will run true to the drilled hole.

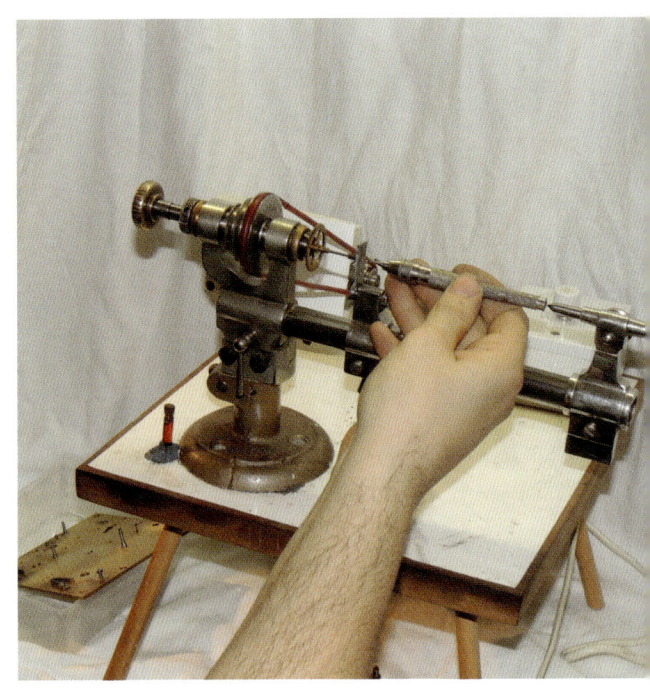

Using the same drill through the hole, drill into the arbor dead on centre.

With the lathe rotating at moderate speed, gently apply pressure to the pin chuck and drill the arbor. It is important to get used to the feel of a cutting drill bit; otherwise you may be tempted to increase pressure if you see no chips pouring from the drill hole. Doing this often results in a broken drill bit which is difficult if not impossible to remove from the arbor. When you feel that the drill is no longer cutting, it most likely needs sharpening. Periodically remove the drill from the hole as you go, wipe the chips off, and lubricate the drill bit before you start drilling again. In softer materials you may see huge amounts of swarf pouring from the pivot hole as drilling progresses at great speed, while in harder materials you may find that you are clearing the flutes of the drill of what appears to be dust. You are still making progress and should periodically check that the drill bit is sharp and lubricated.

Ideally you should drill the arbor to roughly one and a half times to twice as deep as the pivot is long. To measure this, insert the drill as deep as the hole is and hold a fingernail where

Lubricate the drill frequently to help prevent breakage.

The resulting hole will be perfectly true to the arbor.

Remember to thoroughly clean all traces of oil from the new hole.

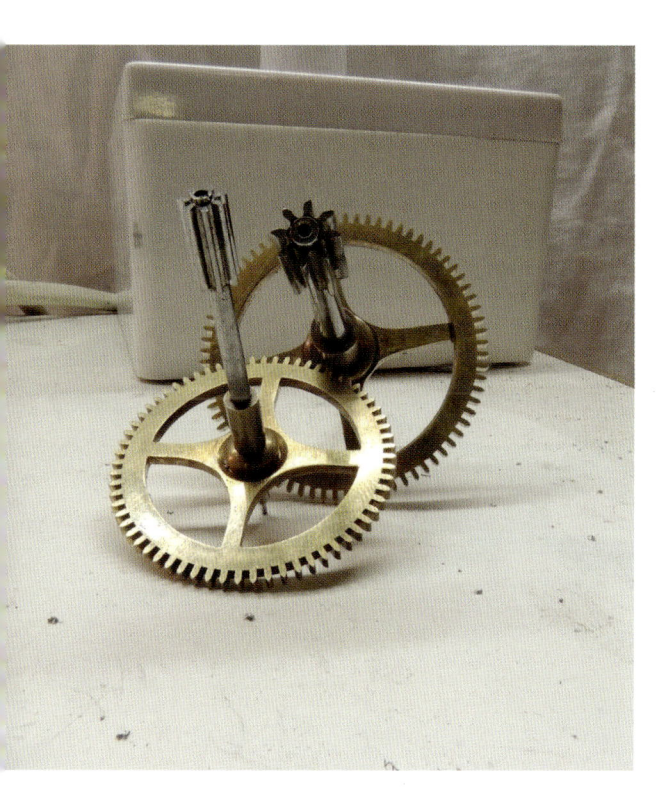

it meets the shoulder, comparing this length to that of the pivot. Occasionally when you are drilling into a pinion, twice the length of the pivot would essentially hollow out the entire pinion, leaving it weakened. In this case I would drill one to one and a half times the depth of the pivot to leave the pinion with a solid core for part of its length. A drilling of less than the length of the pivot would be very weak and likely to become loose in the future.

The hard part of the repair is now complete; you have drilled a dead centre hole in the arbor to a sufficient depth and now all you need to do is insert a pivot. A good selection of blued steel wire, which is available from all horological suppliers, is essential. Select a piece slightly larger in diameter than the hole and mount it in the lathe. Using a graver, face the end of the steel flat and square, and turn down the diameter to a slight taper until the extreme end just starts to enter the hole. Reduce the taper until it is parallel and long enough to fill the drilled hole. Part the steel off at a length roughly twice what is needed to fill the hole and produce the pivot.

Mount the arbor back in the lathe with the hole to be filled facing outwards. Clamp it in tight and, using an oiler, place a tiny amount of Stud-lock onto the hole. I use Stud-lock as a backup to ensure a solid repair and not as the main fixture. The blued steel insert should be a firm fit in the hole, requiring gentle taps from a small hammer to fit. The friction between the two components should be enough to hold the insert in place. If it feels too tight and will not enter with gentle taps, reduce the diameter of the insert slightly with a pivot file and try again. Do not force it or you will split the arbor and ruin it. When the insert is fully driven into the hole, cut the pivot to slightly longer than its final length and stone to finish. Finish the pivot to its final diameter with a pivot file and burnisher, or by turning if the insert diameter is much larger. The drilling guide piece can be used to support the arbor while finishing the pivot.

Turn down a piece of blue steel to just fit the entrance of the drilled hole in the arbor.

With a drop of Stud-lock, tap the arbor onto the blue steel insert.

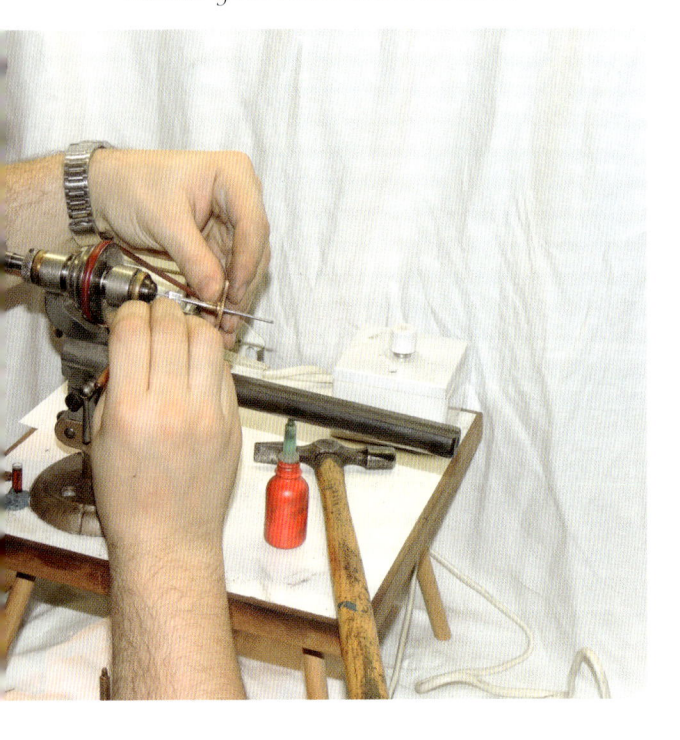

Part the new pivot from the steel using a graver and finish the pivot with the file and burnisher.

The resulting pivot will look and behave just as the original.

Repairs

HOOKING EYE REPAIR

Toolbox 2

The hooking eyes of clock mainsprings are often torn out when the clock is finally submitted for overhaul, and repairing a damaged hooking eye is a basic job which every repairer should master. You need to assess the overall mainspring to be certain that it is not best to just replace it entirely.

With the overall mainspring in good condition and just the outer hooking eye to repair you can begin the work. A good pair of tin snips will cut through most mainsprings, but if the spring is especially thick you can break away the unwanted end by clamping it in the jaws of a bench vice and hammering it over through a sharp angle. Break or cut away the old hooking eye without removing too much length from the mainspring.

Using a diamond file, remove any sharp edges and shape the end of the mainspring to match the original. Temper the end of the mainspring with a blowtorch completely (that is, until it glows a dull red) and when removing the heat be sure to withdraw the flame slowly so as not

STAINLESS OR CARBON?

To test whether a spring is stainless steel or carbon steel, place a small drop of lemon juice on a clean area; carbon steel will react and turn black while stainless steel will remain stainless.

Lemon juice will stain carbon steel but not stainless steel, so use it to identify your metals.

Torn-out hooking eye of an original mainspring.

Cut away the useless part of the spring with tin snips.

Completely anneal the last inch or two of the mainspring; be careful not to let it cool too quickly or air-harden.

Mark with a centre punch where the new hole is to go.

to air-harden the thin steel. Temper the end of the spring to a length relative to its width and the size of the hole required. For example, French springs of 17mm wide require roughly 30–40mm of steel to be tempered, while larger fusee springs will require roughly 40–60mm to be tempered in order to make a sufficient hooking eye, while leaving a sufficient amount of supporting material to remain strong.

You can make the hole in a number of ways, and your choice should be based on your own experience and tooling. During manufacture the hooking eyes are punched out under high pressure. You can replicate this at home using a bench vice and homemade dies; however, you would need several tools of different sizes and the time taken to do this does not really justify the need.

I use the pillar drill and a standard drill bit to make a starter hole. Clamp the mainspring in a vice and support it from underneath with a sacrificial piece of wood which will support the chips as the drill breaks through the thin steel; in my experience this helps reduce the chance of the drill grabbing. Do not hold the spring by hand as you drill; if the drill bit grabs and spins the work piece, the thin steel will easily cut skin. Advance the drill bit slowly and reduce the feed even more as the bit breaks through the final piece of steel.

With a starter hole made, I like to draw the finished shape onto the steel with a marker pen. Use a diamond file to remove material with the spring clamped in a vice, and by hand shape the hole to mimic the original. By filing like this you are able to make small adjustments side to side in order to keep the finished hooking eye in the centre of the spring. Make sure that no sharp angles are left after filing, and if there are, round them in order to reduce stresses and future likelihood of tearing. Use pliers to shape the outer tempered part of the mainspring to a curve which roughly matches the barrel diameter.

If the mainspring is a stainless steel one then tempering and drilling a hole will not work. In this case I suggest you replace the mainspring entirely, or use a grinding stone to grind a hole instead of a drill bit to cut one, and then finish with diamond files as before.

Drill a hole straight through the annealed mainspring.

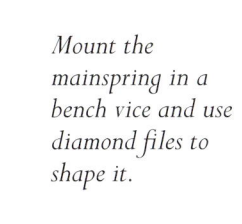

Mount the mainspring in a bench vice and use diamond files to shape it.

Leave no sharp corners (which will easily crack); the resulting hooking eye should be shaped as shown.

NEW MAINSPRING

Toolbox 3

Mainsprings can need replacing for many reasons – tiredness, cracking, rust, torn hooking eyes or deformation, etc.

As a general rule, a mainspring which does not open up to at least three times the diameter of the barrel has effectively 'lost its springiness' and should be replaced. This is because the inner coils of the spring have 'set' and become useless. A set mainspring is usually obvious before you even remove it from the barrel; the inner coils fill the middle of the spring with three or more coils. Replace the spring while the clock is in pieces; attempting to work around it will just result in wasted time and frustration.

Check the inner coils closely for cracking and distortion. This is a common failure point where the mainspring is folded around the barrel arbor hook as the spring is wound.

A little surface rust can be cleaned away with emery cloth, but if there is a significant amount of rust, replace the mainspring; there is no knowing how deep the rust has penetrated and any weak spot is a potential failure point.

Finally, check up and down the spring for any splits, deep scratches and distortion, and check both the inner and outer hooking eyes for damage.

If you have examined and decided to replace a mainspring you will need to order another one from a horological supplier. To do this you will need to take a few measurements, which means you will need to own and be able to read Vernier callipers in millimetres.

You need to take three measurements when ordering a new mainspring: width, thickness and barrel diameter. Start by measuring the width (or height) of the spring; they vary anywhere between 9mm and 28mm usually, but common sizes tend to settle between 15mm and 22mm. Measure to the nearest millimetre.

Next measure the thickness of the spring. Open the spring to allow access for the callipers or micrometre, and while measuring, remember to rock the them side to side until the minimum measurement is obtained; this will be when the jaws are resting flat against the spring and will

Delaminating of steel is not uncommon in older mainsprings and creates a serious weak spot.

Fusee springs have a tendency to break at this point.

Measure the width of the mainspring with Vernier callipers.

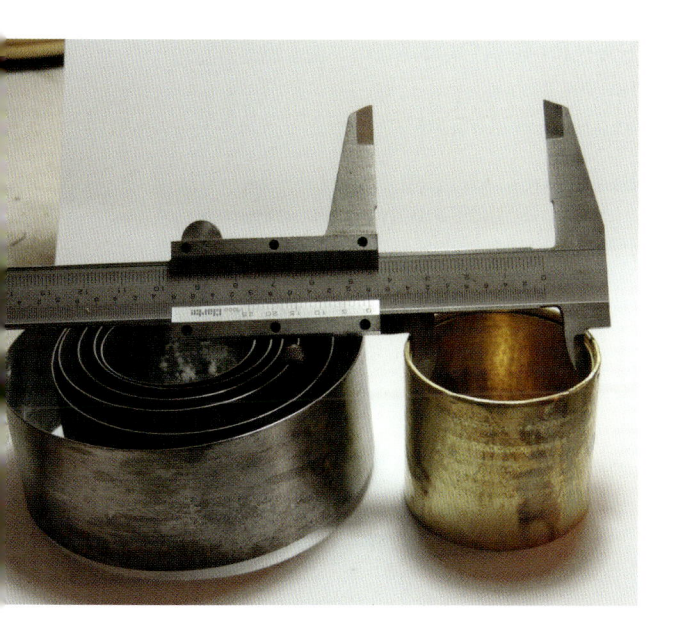

Measure the internal barrel diameter with Vernier callipers.

Measure the thickness of the mainspring with a micrometer.

thus be the true measurement. This tends to be between 0.25mm and 0.5mm, in increments of 0.05mm. Opt for the measurement to the nearest 0.05mm if you do not land on an exact figure.

Finally, you need to measure the barrel diameter; this is to define the length of the spring. Use the inside jaws of your Vernier callipers and take the measurement, and again rock the callipers until they settle at the widest point of the opening. Common sizes are 25mm to 50mm diameter.

What you have now are three numbers – width, thickness and barrel diameter – written, for example, as $19 \times 0.30 \times 30$, meaning 19mm wide by 0.30mm thick to fit a barrel 30mm in diameter. This is how the sizes are expressed by all major suppliers.

A final note on ordering springs is that, if possible, you should opt for carbon steel springs, made the traditional way. Sometimes, however, this is not an option, or it is not mentioned at all. In this situation, some

---

SPRING POWER

We reduce the width rather than the thickness because the relationship between width and strength is direct. In other words, doubling the width will double the strength of a spring, while the relationship between thickness and strength is more complex because it is a function of the cube of the thickness, meaning that if you double the thickness the strength increases $2 \times 2 \times 2 = 8$ times.

For example, if you reduce the width from 19mm to 18mm (one standard size) ($100 \times 18 / 19 = 95$ per cent), you reduce the power output of the mainspring by 5 per cent. If, however, you were to reduce the thickness of the spring from 0.35mm to 0.3mm (one standard size) ($100 \times 0.3 / 0.35 = 85$ per cent), you reduce the thickness by roughly 15 per cent, but you reduce the power output by 37 per cent ($100 \times$ (0.3 cubed) $/$ (0.35 cubed) $= 63$).

Of course, in this example the change from 0.35mm to 0.3mm (15 per cent) is three times larger than the change from 19mm to 18mm (5 per cent), but the change in strength is significant at roughly 7.5 times larger.

suppliers will send you a stainless steel spring, which have a reputation for being stronger than the original carbon springs size for size. If you know a stainless steel spring is inevitable, a good way to get around this problem is to order a spring of 1mm less in width. This will slightly reduce the strength of the stainless spring to match the original more closely, although it is in no way an exact result.

BARREL HOOKS

Toolbox 3

Barrel hooks are under constant strain from the mainspring, and any power fluctuations in the spring, such as a breakage or tight winding, can put enough force on the barrel hook to tear it from the barrel.

Measure the depth of the barrel from the cap groove using a depth gauge.

Measure the depth of the barrel from the very top.

Barrel hooks come in two forms, one being the cheaper sheared brass hook, where a high pressure tool is used to force the brass barrel to shear inward and form the hook. The other form used in antique clocks has a steel hook screwed and riveted into the barrel. When either form of barrel hook becomes damaged, it should be replaced with the latter type, which is far more practical to make in the workshop.

If the original barrel hook hole is a mess, as it usually is, especially if the sheared brass hook is being repaired, then you will be making a new hole 180 degrees away at the opposite side of the barrel. If the original hole is intact, you can simply re-thread it.

If a new hole is needed, it will need to be accurately placed at the centre of the mainspring's width, not the centre of the barrel

Mark the outside of the barrel with the calculated centre height of the mainspring.

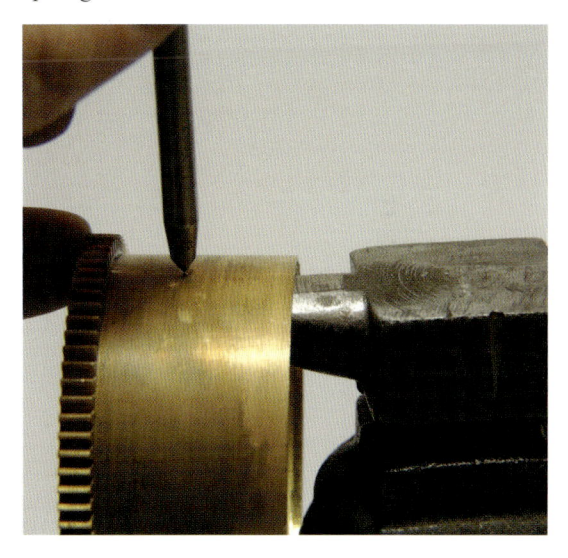

Centre-punch a mark at the centre height.

Drill a through hole to be tapped and take the new barrel hook.

New barrel hook fitted but not filed to shape; mark it up for orientation before filing.

width. To find this point, use the depth gauge at the end of your Vernier callipers to measure the depth of the barrel from the groove in which the barrel cap fits. Halve this value and add to it the depth of the barrel cap groove. Set the callipers to this new value and use them to scribe a mark representing the correct height of the barrel hook. Take a good sharp centre punch and make a light indentation to guide the drill bit. Use a drill undersized for the final diameter of thread to be used and, with the barrel well supported in either in a V-block or a machine vice, drill the hole.

Select a steel cheese-head or countersunk machine-type screw of a diameter to roughly match that of the original hook (or its hole). Broach the hole in the barrel until it starts to accept the correct tap to match the screw thread, and tap the barrel.

Insert the machine screw from the inside of the barrel until it will go no further, and use a marker pen to mark off the excess thread and the front face of the cheese-head which will form the hook. Remove the screw and cut off the excess thread using a junior hacksaw or piercing saw, leaving enough to form a rivet. Before you refit the screw, file a chamfer to the underside of the cheese-head screw where it was marked to form the hook, and reduce the height of the head to a reasonable height. The hook should protrude just enough to hook the mainspring and draw it against the inside of the barrel.

Use a large drill bit to lightly chamfer the outer edge of the screw hole; this is to take the rivet. Refit the screw until the chamfer faces in the correct direction, which is usually to the left if viewed through the hole, although this is not a certainty. Finally, resting the head of the screw on an anvil, use a ball-pein hammer to rivet over the remaining thread into the chamfer. When the job is finished all that remains is to file away the protruding rivet and refinish the barrel.

Finished barrel hook has been left taller than preferable for clarity of the photograph.

Externally the barrel hook should look as though it has always been there.

WINDING SQUARE

Toolbox 1

Winding squares often become rounded over when the clock's owner uses a poorly fitting winding key. When you first see a badly rounded winding square, check the key for fit, and if necessary supply a smaller key to the owner which will stop this damage from recurring in the future, but before fitting the new key, you need to repair the winding square.

It is best by far to leave as much of the original material in place as possible. This means that, rather than filing a damaged square back to shape and reducing its size, it is preferable to rivet over the burrs at the corners of the square and hammer it back to shape. This allows the square to retain its strength, and in many cases for the original winding key to remain in use.

Hold the winding arbor with the square resting on a flat anvil and lightly tap along the edge with the flat face of a hammer; as you approach the pivot, to ensure you do not damage it, switch to using a flat-faced punch. Repeat this step for all four edges and test-fit the key. If the corners are sharp after this treatment, quickly burnish them with an oval burnisher to flatten any remains of burr.

In serious cases it may be unavoidable and you will have to file a new square (I have seen winding squares near perfectly round). To file a square, clamp the arbor in the vice with the surface to be filed protruding and file a flat; do not file too deep, just enough to register as a datum point. Rotate the part in the vice so that the flat registers against one of the jaws, file another flat on the protruding surface. By keeping the file at the same angle, you have now produced two flats at 90 degrees to each other. Continue to rotate the part in the vice by 90 degrees, and file away a small amount of

This winding square has seen better days and is probably the result of a poorly fitted winding key.

Hammer the square on an anvil to reduce the burrs and restore it to shape.

material from the uppermost surface until the flats almost meet at the corner. At this point it is time to start fitting a key. Try an appropriate key on the square; it should fit on easily, but have minimal freedom. Continue to reduce the square until it is the perfect fit for the chosen key. If you have filed until the flats meet, burnish the corner to reduce the sharpness.

If you have two or more winding squares, it is best not to reduce the perfectly good square to match the repaired one. In this situation, discuss with the owner whether to go to the added expense of making or sourcing a new arbor completely, or whether to supply a second winding key. Never introduce a new problem to mask an old one.

The result of hammering the square can be quite satisfactory.

When tidied up with light filing and polishing, the finished barrel square looks almost as good as new.

To file a square, mount the round stock in a vice and file the top flat.

Keep the file square to the jaws and the resulting flat will be near-perfect.

Rotate the stock in the vice to register the flat against one jaw; file a new flat at the same angle as before.

The result is a near-perfect 90-degree angle; repeat the process with all four sides for a perfect square.

RATCHET WHEEL

Toolbox 1

The ratchet teeth can become chewed up or worn over time, especially if the click spring breaks or the click screw comes loose. Occasionally you can tidy up the ratchet wheel with gentle tooth manipulation and some filing; however, often it is best to replace the wheel entirely.

If the ratchet wheel needs replacing you will have to find a suitable source. You can either get one cut (which means measuring all dimensions and counting the teeth so that you can place an order with your preferred supplier) or you may be able to buy a pre-cut blank from the supply house. Alternatively, you can fit one from a scrap movement of a similar kind.

To select the right ratchet wheel from a scrap movement, you will want to find one with the correct diameter and of a similar thickness. Some modification of the centre square will most probably be needed, and this can be done with your selection of needle files. Do not be tempted to file down the winding square to fit the new ratchet wheel.

CLICK AND SCREW

Toolbox 3

It is not uncommon for the thread of the click screw or its hole to become ruined over time. It is one of the most important screws in the clock and must not be overlooked. There are several ways to fix ruined thread on a click screw or hole.

If the hole itself is damaged and the click screw is in good order, then you can broach out and bush the screw hole, providing fresh brass into which you can cut a new thread. If you take this approach it will be important to rivet the bush in place to ensure that the hold is solid, otherwise it is no better than a ruined thread.

If the click screw has a ruined thread, then it is best to make or source a new one. A lot of time can be saved if a salvaged click screw can be modified to fit the existing click. To do this, select a click screw which is tall enough to accept the click – oversized is fine as you will be modifying it in the lathe. The thread should be the correct size for the existing hole, or slightly larger so that a new thread can be cut. Put the screw in the lathe and trim the diameter of the shoulder until the click is a good fit and free to move around, then mark and trim it to length. Finally, select a die of the correct size and continue the thread to meet the new shoulder. Should the new thread be too long, trim the end so that it does not protrude into and foul the components between the plates.

If the new screw has a larger thread than the hole, simply enlarge the hole with the correct size of tap to suit.

If you have no spare screws to modify, it only takes a little longer to make the screw from blued steel, and is essentially a turning exercise in the lathe. Style the screw to match the original and re-blue it when you are happy with the fit.

Should the click itself become damaged then it is usually at the tail end, which is not a functioning component and can be left if it is not decorative. If the head of the click is damaged beyond a few raised burrs, then a new part will need to be sourced, otherwise the burrs can be filed away and the click blued.

Ruined click screw (right) with a potential donor screw (left).

The shoulder of the donor square is turned down to fit the click.

The shoulder remains slightly too tall and should be reduced in height.

Chase the thread with an appropriate die up until the shoulder.

Refinish the screw to match the original in finish.

Resulting screw is a perfect fit for the click – a safe and long-lasting repair.

Chasing the thread in a screw plate.

CLICK SPRING

Toolbox 2

Click springs often break but can be an easy fix. For the majority of modern clocks (less than 100 years old) you can buy springs of the correct size and shape, for longcase clocks you can buy blanks and for many others they are simple enough to make from spring steel filed to shape, or a piece of brass using the old spring as a template.

If the clock is French it is best to have a supply of scrap movements as many of the components are interchangeable or will swap with little modification. Boxes of scrap movements at auction are usually very affordable.

All parts can be made from stock material and old mainsprings are especially useful for making steel click springs. If the original is available then it will make a handy template. I like to attach the original pieces to some stock material with a drop of superglue so I can assemble a good template. When the glue is set, give it a quick puff of paint from a spray can, break away the superglue joint and you have a perfect template which you can cut and file to match the original – even the steady pin and screw holes will be marked.

With the template cut and filed to shape you need to work-harden the brass or harden and temper the steel. To work-harden the brass, hammer it gently all over, enough to distort the finish but not so much as to ruin the piece, then file to final dimensions and finish as necessary. Steel parts are finished before hardening and tempering; when you are happy with the look of the part it is time to harden it. When the part is tempered to dark blue, refinish it to the required standard and if necessary blue it with heat for aesthetic effect.

BENT TEETH

Toolbox 1

Bent teeth can often be straightened by hand. Clearly a bent tooth will not allow the uninterrupted passage of the pinion as the pair rotate together, and unevenly separated teeth will cause excess friction or butting.

A bent tooth on a train wheel is not an uncommon sight.

Be sure to get the screwdriver blade into the root of the next tooth before prying.

To straighten a tooth it is important to inspect the roots for any cracking. Teeth that are cracked or bent through more than 45 degrees rarely if ever survive being straightened, so prepare yourself to deal with broken teeth if this is the case.

To straighten a bent tooth use a sturdy screwdriver which will fit between the tooth gaps. Your ten-piece watchmaker's screwdriver set is perfect for this. Place the blade of the screwdriver in against the very base of the next tooth, and lever against it. If you do this too high up the opposite tooth then all you will do is damage that one too, so be careful to keep the screwdriver firmly against the root as you lever the bent tooth back into position. When the tooth appears straight, and the tooth gaps either side are even, inspect the root for cracks, as a cracked tooth will bend or fall right off.

Toolbox 3

Escape wheel teeth are best straightened with a pair of smooth-jawed pliers. Grip the bent tooth between the jaws and lightly pull with the pliers, allowing the tooth to slip in their grip until it is released. This method will eventually straighten the tooth with minimum likelihood of breaking the tip off; however, it is possible for the tooth to be stretched during this operation and therefore be too long, causing running problems.

Escape teeth which have been straightened in this way may need to be topped. Topping wheels is a very delicate operation and should be done with care. Mount the escape wheel in the lathe and set it running; gently raise the edge of a piece of paper to lightly touch the spinning wheel. A long tooth or eccentric wheel can be heard: one long tooth will catch the edge of the paper with unequal depth to all other teeth, and the buzzing noise will be uneven. If you suspect a long tooth, it is wise to lightly touch the spinning wheel with a stone or fine diamond lap until it shows a slight hint of brass on its surface. Be very careful to touch the spinning wheel very lightly with the stone; you are only trying to shorten the one long tooth, and doing so too aggressively will bend it all over again.

Bent escape teeth can be more troublesome than train teeth but are just as common.

Pull the bent tooth gently with smooth-jawed pliers to straighten it.

Testing for short or long teeth by listening for changes in the vibration of a piece of paper resting against the teeth.

Topping the escape wheel to reduce the tall tooth by gently resting a diamond lap against the wheel as it spins.

TOOTH WEAR

Toolbox 3

When wheel teeth wear, their carefully designed profiles are ruined and the freedom between wheel and pinion increases dramatically. The result is excessive friction and a clock that will not run reliably. Clock wheels turn in one direction only and so only wear on one side. This allows you the opportunity in some cases to remount the wheel the other way around, increasing its life, maintaining originality and resulting in a more affordable repair.

Teeth wear at the addendum, so the strength of the tooth is not compromised. Provided the depth of wear is no more than roughly one third of its width, you are able to safely turn the wheel around to achieve another 200 years of running from the original part. If, however, the wear is as deep as halfway into the width of the tooth (as I have seen on more than one occasion), the result of turning a wheel around would be poor and it should be replaced with a new one.

Removing the wheel depends on how it is mounted; most are riveted either to a collet or directly to a pinion. If the wheel is riveted to a brass collet then it is possible to punch

Worn train teeth are not uncommon; this example is a very mild case.

Turning the wheel over will introduce unworn tooth surface to the pinion, improving the situation and preserving originality.

enough to hold tight when fitted). Next, on the other side of the broken teeth, cut once again into the tooth gap, at an angle equal but opposite to the angle just cut. Join these two cuts together by sawing in a straight line between them; you will be left with a dovetail-shaped gap.

The reason you dovetail the insert is to reduce the chance of it coming loose with time. If you drill or saw a straight slot into which you insert a new tooth, the lateral driving pressures to which the tooth is subject will eventually cause it to wriggle free and fall out (in the same way as you would work a fence post out of the ground), but a dovetailed tooth will be held in place by its

A dovetailed cut-out is the beginning of a strong repair and worth the additional time in getting it right.

Finding an exact match donor wheel is unlikely, but you should spend time finding the closest match possible to reduce the likelihood of division problems.

Repairs 151

shape, which spreads the lateral forces throughout the rim of the wheel without loosening the opening hole.

Now cut a section from the scrap wheel to match the dovetail in your repaired wheel. My only advice here is to cut it oversized using the piercing saw. Then, grasping the small insert in a pin vice, file it to shape until it is a good tight fit. Use a needle file with a safe edge to avoid damaging the tooth.

When the insert fits the hole, smear a little flux along its filed edges and press it into place on an anvil using a flat-faced punch and finger pressure; this will ensure it fits flush with the

A good dovetail will be a tight fit which will hold together on its own; the soft solder only supports the repair.

As with all good repairs, you should have to look closely to spot it when finished.

wheel. Place a small slither of soft solder on each lower corner of the dovetail; the corners are cut with a piercing saw and likely to have a slight gap which will allow the solder to flow into and around the edge of the insert, holding it firmly in place. Warm the wheel with a blowtorch until the solder flows easily. When everything has cooled down, remove any excess solder and grain or polish the wheel to match the original.

If the teeth of the insert are slightly long or wide, file them to shape using a thin needle file which has fine teeth on only one face so as not to damage surrounding teeth. With a well-selected scrap wheel, this should be unnecessary.

It is possible that the insert could be a brass blank which, after soldering into position, can have teeth cut into it. This ensures the perfect separation and shape of teeth, but requires wheel-cutting equipment.

Lateral forces on a tooth which is not dovetailed will soon work it free.

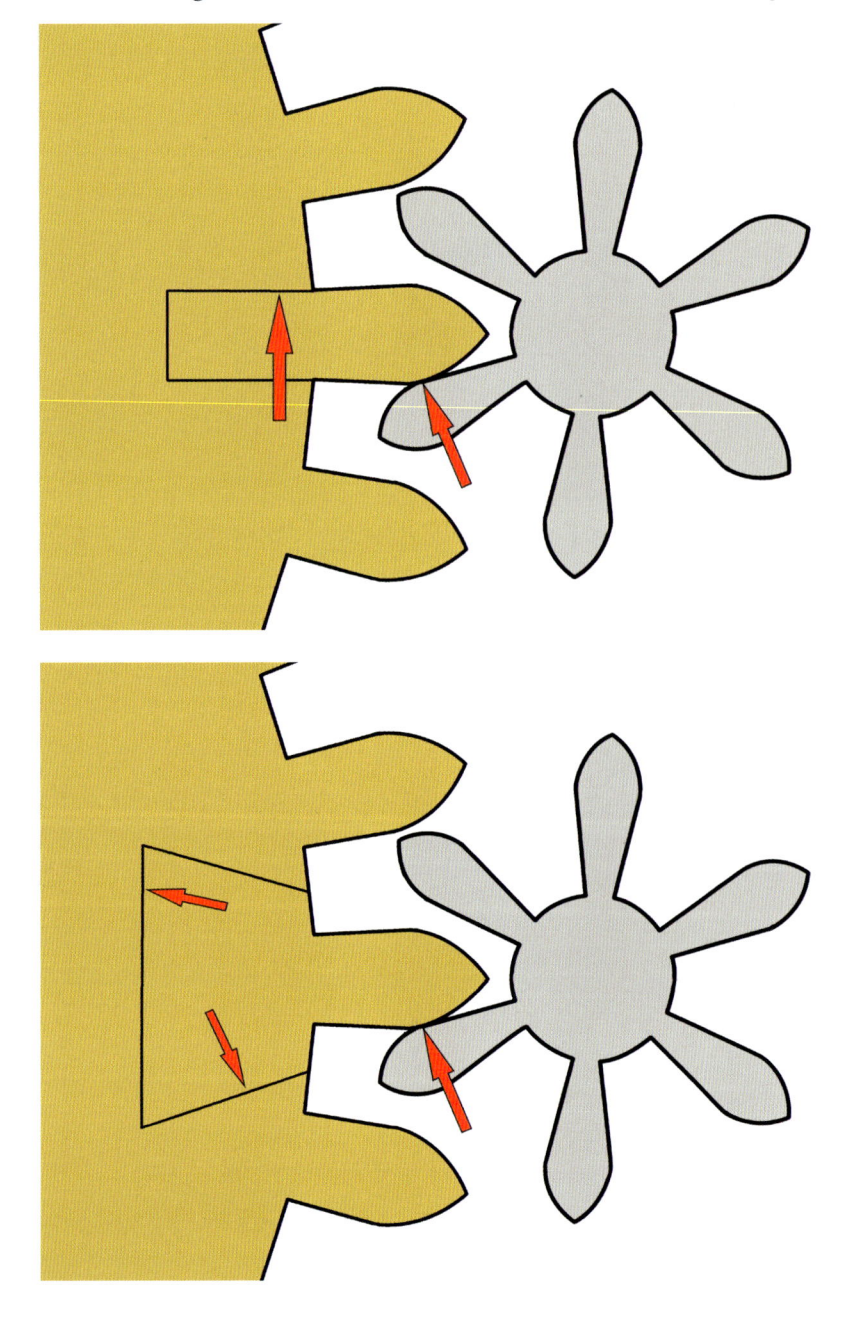

Lateral forces on a dovetailed tooth repair are spread throughout the rim of the wheel, making it much stronger.

BARREL TEETH

Toolbox 3

Barrel teeth are repaired similarly to the teeth of any wheel, by dovetailing in a section from a scrap barrel of similar dimensions. However, because of the design of the barrel, you are not able to simply pierce out a section.

Instead, the repair involves drilling a starter hole, piercing out the dovetail, and de-soldering the piece from the barrel. The process is best explained by the photographs provided. With the damaged barrel section removed, you heat the donor barrel to remove the gear as shown, and continue to pierce a suitable dovetail with which to repair your barrel. Finish by soldering the dovetailed piece into the barrel with soft solder.

Broken barrel teeth are another common result of spring failure; with the starter holes drilled, use the piercing saw to cut the dovetail shape (in red).

Do not saw along this line; most barrel gears are soldered to the barrel and cutting here serves only to weaken the joint.

De-solder the unwanted piece of barrel gear; the new section can then be soldered back to the remaining shoulder, forming a much stronger connection.

The only reason to differ from this repair is if the barrel has been gilded, which is common for modern carriage clocks, or if the barrel is made from a solid piece. Heating, de-soldering, soldering and so on of these barrels would ruin the finish and make for an ugly unprofessional repair. Unless you intend to have it re-gilded upon completion, the following repair, whilst not as strong, will suffice for all but the strongest of mainsprings.

Using a file, remove what is left of the offending tooth, whilst leaving a slight trace of its root to identify where the new tooth is to be placed.

Select a drill bit of a diameter matching the width of the tooth, and mount it in the drill press. The depth of a barrel tooth is often four or more times the width of the tooth. By drilling and inserting pins you are able to disperse the load, often between three or more pins, resulting in a repair which is strong enough to

The majority of barrel gears are soft-soldered onto their barrels and can be disassembled as shown.

On gilded barrels an alternative method is needed to preserve the finish; drill the root of the tooth to take inserts.

The brass pins which will make the new tooth should be driven well into their holes using a little Studlock to ensure their grip.

be accepted where dovetailing is not suitable.

Drill a series of holes spaced evenly along the witness mark of the original tooth. Be sure to leave enough material between the drillings that they do not burst through when a pin is inserted.

Use a fine oiler to place a tiny drop of Studlock into each of these holes, and insert a brass taper pin. Use a hammer to knock in the taper pins until they are tight. Cut off most of the excess length with a pair of cutters, and use a needle file to shape the teeth to the correct form.

All that is left to do is to remove the excess material from inside the barrel. Most of this can be cut away with side cutters, and filed flush to the barrel (so as not to impede the mainspring, which would press the insert out from beneath as it unwinds). I use the Dremel tool with a sanding drum attachment, which is much quicker.

The brass pins which make the new tooth should not protrude into the barrel where the mainspring will be.

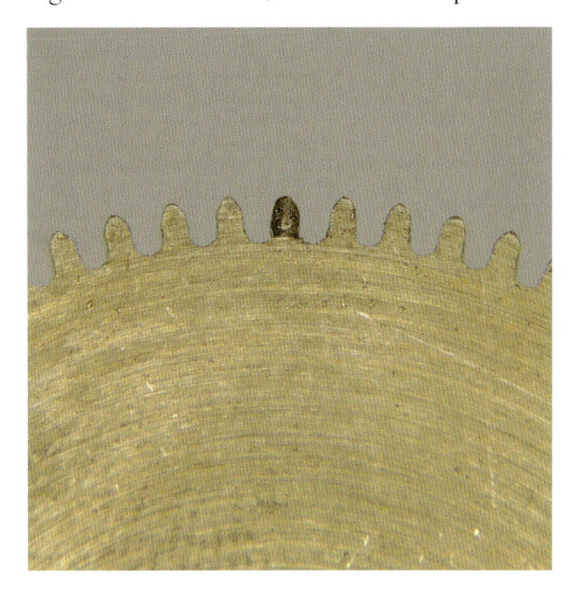

When filed to shape, the height and profile of the insert should be a close match to the original teeth.

The resulting tooth repair is an imperfect but acceptable one in many situations and should last a number of years.

FITTING A NEW WHEEL

Toolbox 3

For replacing the wheel entirely, you will need to contact one of the many specialists who offer the service of wheel cutting. Many of these will accept a wheel and arbor by post and return to you a complete job, ready to mount back between the plates; teeth cut, wheel crossed and fitted to the pinion or collet. I prefer to buy blank wheels cut with teeth, but not crossed or mounted. They can be ordered by email by quoting the diameter across the tooth tips, number of teeth, type of tooth form (train, recoil, dead-beat, etc.) and thickness of the wheel. When the blank arrives by post, cross and mount it to the arbor, which should take no more than thirty minutes. If you choose this route, you must work accurately, as an eccentric wheel cannot be blamed on the supplier if you have chosen to mount it yourself.

The moment the wheel arrives, double-check all dimensions by measuring the wheel in comparison to the original and counting the teeth. The thickness is not so important that small variations cannot be accepted.

Start by crossing out the wheel or cutting

New wheel blank as it arrived from the supplier.

Marking the crossings by using the old wheel as a guide.

the spokes. My preferred method is to hold the original wheel on top of the new wheel and scribe around the crossings as closely as you can. If the original wheel is missing, then you will need to scribe lines directly onto the wheel with a ruler and a pair of compasses. Make sure that the number, design and taper of the spokes, including their final thicknesses, matches the original. It is best to draw the design to scale on paper first; when you are happy, use contact adhesive to glue the paper to the wheel. The paper can be drilled and cut without problem.

Use a small drill to make holes in the corners of each crossing, getting as close as you can to the line without going over it; a drill press is useful here because it allows you to focus on placement of the work piece.

With the holes made, select a piercing saw blade to allow roughly three teeth within the thickness of the material at any time, and a piercing saw frame which will clear the wheel on the deepest cuts. Thread the saw blade through a hole so that the teeth will cut on the pull stroke, and attach the saw frame. Cutting as close as you can to the guide lines without distorting them saves a lot of time in finishing, and with practice you will be able to saw right along these lines, leaving only saw marks

Drilling the corners of the crossing; these holes will then be connected with the piercing saw.

Cut close to the line and finish with files and stones.

to be filed away. If you are not confident with the piercing saw, cut half a millimetre or so away from the line. Join together all drilled holes until the centre of the crossing falls free. File to shape using needle files, some of which should be modified by grinding to present a polished edge to the surface which is not to be cut. When filing curved inside surfaces, it is easiest to use a half-round file, and to roll the wheel back and forth on the workbench as you file; this helps to avoid pits and create a continuous surface. File into the corners using a flat or curved file (depending on whether you are filing a flat or curved surface) with a safe edge; this will stop you from cutting into the resting surface and allow a crisp sharp corner. Finish to a standard which equals the original wheels. There is no need for a perfect finish on mass-produced clocks whose wheels were sheared or cast and left unfinished.

With the wheel crossed out, all that is left to do is mount it to the arbor. Prepare the lathe by

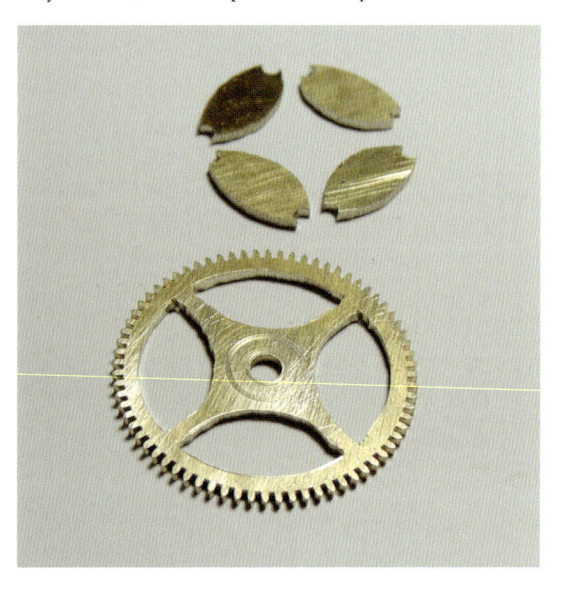

New wheel with the crossing roughed out.

A cutting broach will open the centre hole concentrically, unlike a drill which may wander.

The wheel should be a tight fit onto its arbor; use a hammer to make sure it rests fully against the shoulder.

mounting the arbor and checking it runs true. Turn the seat onto which the wheel mounts until it runs true. If there is no excess material to form a rivet, cut into the pinion or collet to create a slight excess in length. Using a five-sided cutting broach, open the hole in the wheel until it is a tight fit on the collet or pinion. A broach will cut the hole concentrically, unlike a drill which may wander. Press the wheel into place with the side which the broach entered from facing outwards; this is so that the taper opens out towards the rivet and produces the chamfer for the rivet to spread into. Now mount the wheel on an anvil to support the collet or pinion from beneath. Rivet the collet or pinion leaves and not the wheel; they will spread to clamp the wheel in place.

Check for concentricity and that the wheel runs true in the lathe. Finish to the standard required for the clock: the example shown came from a mass-produced clock and so was not given a perfect finish.

Rivet the wheel into place using a domed then a flat-faced punch.

Finish the wheel to the standard required for that clock.

PINION WEAR

Toolbox 2

It is common to see the steel pinion leaves of a clock cut with deep grooves by the action of the brass wheel sliding over their surface. These grooves can become surprisingly deep before they become troublesome in some cases, but any sign of wear on the pinion leaves is evidence of increased friction in the train, and potentially the reason for a troublesome clock.

To repair this you either replace the material in the pinion using micro-welding, or move the wheel along the arbor until the wheel is running on an untouched pinion surface. The worst examples of this repair I have seen had the wheel dished in order to move the teeth sideways relative to the wheel hub. I have never seen a dished wheel which runs true, and I have seen plenty with cracked and soldered spokes. Other repairers heat the collet to soften the solder, so that the collet can slide along the arbor by the necessary amount. This is closer to a good repair, but often results in an eccentric wheel (as explained above).

The best way to move a colleted wheel along its arbor without affecting its concentricity is not to soften the solder, allowing the collet to sag, but to shear the solder, so that it maintains its original thickness around the joint. This involves putting pressure along the axis of the arbor, and a tight collet may cause it to bow if your attention lapses.

You will need access to a large bench vice and a selection of anvils, or a gauge plate to make anvils from. Select or make an anvil which will fit snugly over the arbor and rest against the back of the wheel collet, which is deep enough that the pivot does not protrude from the end. The pivot should be contained within the anvil by a depth of at least 5mm to allow for movement.

Select an anvil which fits snugly against the arbor from the opposite end, and will rest at the root of the pinion, with the pivot covered by at least 1mm to keep it protected. Using a marker pen, draw a circle around the end of the collet at the end towards which you will move

Result of moving a wheel along its arbor; note how the wheel no longer meshes with the worn pinion.

Set-up for pressing a wheel along its arbor; make sure the pivots are well protected before attempting this.

the wheel. This mark will define the distance you move the collet, as well as provide visible evidence that the wheel is actually moving.

Set the wheel up in the jaws of the vice so that the anvils are pinched between them and the arbor spans the gap between the anvils. The arbor should be square in all planes and be supported by the pinion, and the soldered collet pinched between the anvils. As you tighten the jaws of the vice, the soldered joint between collet and arbor should begin to shear, and the wheel and collet will slide along the arbor. Keep closing the vice until the penned line in front of the collet disappears beneath it. The collet can be moved over by a very accurate amount when using this method; however, there is a risk of distorting the arbor. As you begin to apply pressure, watch carefully from all angles for any bowing: if it is slight, apply thumb pressure at the peak of distortion in order to counteract it; sometimes this is enough to stop the bowing and the wheel will slide across. If this does not work, stop and apply a little heat

Sheared solder joint resulting from pressing a collet along its arbor.

to the collet, not so much that the solder melts, but if heated gradually you will find a point at which the solder weakens enough to shear without becoming liquid.

If you cannot move the collet, which is not uncommon, the best bet is to dismount the wheel and turn away a portion of the collet shoulder before refitting the wheel in this new position. This is explained in the previous section.

If the wheel you intend to move is riveted to its own pinion, there is no option but to dismount the wheel and modify its seating. First, with all of the wheels mounted between the plates, check that there is clearance to move the wheel deeper into the pinion. Often, such as in the cen-

Mounting a wheel in the lathe to turn away the rivet.

Rivet has now been removed, leaving as much material in place as possible.

tre wheel of longcase clocks, there is very little clearance and moving the wheel would cause it to catch on the screw which holds the key piece onto the greatwheel. If this is the case then it is best to send the arbor to a specialist for micro-welding. If, however, there is plenty of room for moving the wheel over, the following procedure should serve as a guide in most cases.

Fit the arbor into the lathe as close as possible to the pinion. With the lathe running at moderate speed, turn away the tips of the pinion leaves which make the rivet that holds the wheel on. With the rivet cut away, it should be possible to wriggle the wheel free of the arbor, but before doing so, mark a pinion leaf and a corresponding point on the wheel to ensure it

With the rivet turned away in the lathe, it is safe to punch the arbor out of the wheel centre as with a brass-colleted wheel.

Pinion mounted in the lathe in preparation for having a new seat cut.

New seat is cut into the pinion by just over the thickness of the wheel.

will be remounted in the same orientation.

It is possible to turn away the area of pinion which you need to remove, but the knocking cut damages graver tips and it is possible to snap off a leaf if you slip. I prefer to use a pivot file, with its safe edge resting against the centre on which the wheel is mounted and the cutting surface acting on the seat. This way you are guaranteed a sharp corner, no damage to the centre and, most importantly, no damage to the pinion leaves. With the lathe running at moderate speed, work the file back and forth until a wheel's width of material has been removed from the seat. Mount the wheel temporarily to check its fit, and try it between the plates with its partner pinion to ensure enough material has been removed. Remove more material if needed. With the arbor mounted back in the lathe and the wheel pushed into place, turn away the excess pinion leaving just enough to form a rivet. Mount the arbor on an anvil and rivet the pinion over to mount the wheel.

Excess material should now be turned away, leaving just enough to make a new rivet.

Correct amount of material left proud of the wheel in order to form a strong rivet.

Use a punch and hammer to form the rivet which holds the wheel in place; punch marks can be removed in the lathe later.

Finished wheel is now seated a full tooth thickness further down its arbor.

LANTERN PINIONS

Toolbox 2

Lantern pinions wear like any other pinion leaf and easily snap when a mainspring goes bang, conveniently acting in a way similar to a circuit breaker and saving damage to the wheel teeth which are much harder to repair. However the damage presents itself, the repair is the same.

We need to remove the broken, bent or worn steel trundles and the easiest way to do this is to cut them in half with a pair of strong side cutters and pull them out. You will notice that the brass end-caps have holes for each trundle; one cap has blind holes and the other has through holes which have been riveted to hold the trundle in place. If one of these caps has twisted, perhaps with the force of a broken mainspring, simply twist it back into position with a pair of combination pliers, and put a drop of Stud-lock at its centre. Broach the riveted through holes with a smoothing broach to open the rivet for the insertion of standard blue steel rod of the same diameter as the original trundles. Insert the steel rod until it reaches the bottom of the blind hole in the opposite cap. Mark it where it exits the hole and cut it just below this point. Cut enough trundles to this length to fill the remaining holes.

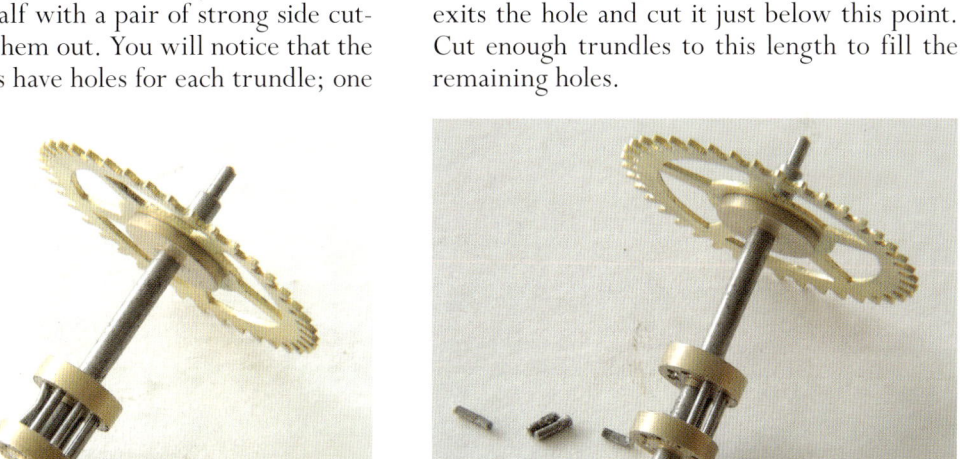

Bent trundles are another common result of spring failure, but they are an easy fix which only takes a few minutes.

Cut the damaged trundles in half to aid in removing them; I used the piercing saw to do this but side cutters usually work too.

Use a smoothing broach to open the hole in the trundle just enough to allow the entrance of a piece of blued steel of the desired diameter.

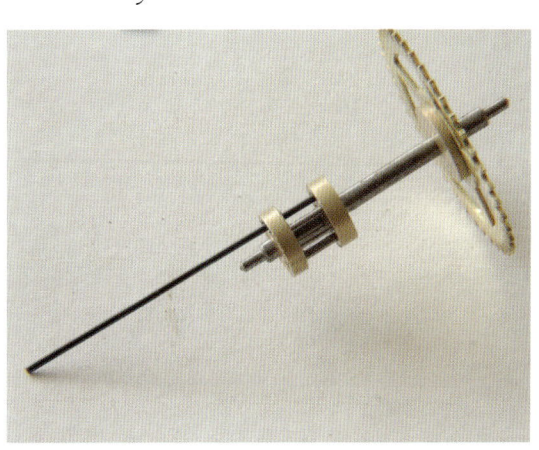

Mark the blued steel insert to a length just shorter than the full length of the trundle.

Find a way to support the top end-cap in such a way that it can be riveted without sliding down the arbor as you hammer it. (Cut a V out of a piece of steel plate thin enough to fit between the two end-caps; clamp this in the bench vice as an anvil.) Place one of the trundles in position and, using a sharp punch, close the end of the through hole to trap the trundle in place. It does not matter if the trundle is free to rotate, so long as it does not wobble side to side. Repeat this process 180 degrees away in the opposite trundle hole, then slowly add more trundles until you can no longer support the top cap in the steel V anvil. At this point the trundles themselves can support the top cap for riveting, so support the bottom cap from beneath on the steel V-anvil. Repeat this step until all of the trundles have been replaced.

Cut enough trundles to this length to complete the repair.

Place the new trundles in their holes and rivet the brass end-cap back down to trap them in place.

Lantern pinion with a single trundle riveted back into place.

Completed trundle repair; to remove the blue colour, a brief soak in rust remover will suffice.

RACK-TAILS

Toolbox 3

Repairing rack-tails is a common enough occurrence for it to be included here, although it is not exactly a beginner's job. Some further elements of strike theory are covered here; they were intentionally omitted from the main theory sections because they are not necessary knowledge until you repair your first rack-tail.

If for any reason the strike fails to run, the face of the snail will crash into the rack-tail between twelve o'clock and one o'clock, causing one of the following problems:

- the hour hand will fail to move but the clock will continue to run
- the rack-tail will 'ride' over the snail as they should be designed to do on English clocks, and although the clock will continue to run, the strike will not
- the rack-tail will be forced backwards as the snail proceeds to bend or break it.

For the first two situations you can correct the problem by winding the clock and turning the hands backwards by twenty minutes or so, which will release the rack-tail from the face of the snail, freeing the strike to run.

If, however, the rack-tail has been broken, you will need to manufacture and fit a new one. When making a new rack-tail you need to consider the following points:

- The relative lengths of the rack (from pivot to teeth) and the rack-tail mean that the two act as a pair of levers would, with the movement of the rack tail being exaggerated in the movement of the rack.
- This ratio is important when replacing the rack-tail as any alterations here will result in mis-gathering of the teeth.
- The ratio of the tooth gap, from first to twelfth tooth, and the height of the step in the snail (which represents eleven teeth) should equal the rack to rack-tail ratio.
- If the rack-tail is missing, or too damaged to be measurable, you can calculate its length using these ratios.

To find the length of a rack-tail, use callipers to measure the distance between tooth one and tooth twelve on the rack, and divide this num-

The ratios of the two red components must equal the ratio of the two green components when making a new rack-tail if it is to work correctly.

ber by twelve. Next, use a depth gauge to measure the height of the step in the snail, and divide this number by eleven. Divide the first answer by the second to get the ratio.

Next, measure the rack from pivot to tooth tip, and multiply the answer by the ratio given. This should equal the length of the rack-tail from pivot point to rack pin.

Start the procedure by removing the old rack-tail from the rack. Often these have been soldered to stop them from slipping, but originally they were riveted. I find the best way to do this job quickly is to grip the rack-tail in pliers and roll them towards the joint, which will peel away the rack-tail.

Clean up what remains of the old rivet, and heat and scrub away with a wire brush any solder which may be present.

Procure a piece of 1mm brass sheet, The brass should be 'hard' so you can take advantage of its springy properties, but if it is not, work-harden it by hammering now. Drill a hole at one end and broach it to fit the rivet, then fit it loosely in place. Mark the point at which the rack pin needs to be fitted by using the calculation above. Drill a small hole here as a marker.

Remove the rack from the piece of brass and cut it to shape. Tin snips will make quick work of it but final shaping will require filing.

To make the rack pin I like to use a shoulder screw. Broach the guide hole to accept a suitable tap and cut a thread. In the lathe, turn away the head of the screw and reduce the shoulder to a suitable diameter to make the pin: 2–3mm should do it. Screw the pin into the rack-tail and mark the face which will be first to come into contact with the step of the snail if the strike fails. Remove the pin and hold it in a pin chuck. While resting against a wooden block, file a 45-degree angle onto the marked face so that it will have the tendency to climb over the snail in the event of a collision.

Screw the pin back into the rack-tail and file away the excess thread, leaving a small amount with which you can form a rivet. Using the ball of a hammer, or a domed punch, rest the pin on a steel anvil and rivet it in place.

Fit the rack-tail to the rack and lightly rivet it in place using a domed punch. Fit the rack to the clock and set it for one o'clock. Adjust the angle of the rack to rack-tail until the rack-hook rests perfectly in the first tooth gap, move on to two o clock, then three, making minute adjustments until the best compromise is reached. The rack-hook should rest nicely in the tooth gap for each step of the snail; however, there is some tolerance built into the design. Complete the rivet until it is firm. If you have any doubts reinforce it by drilling a hole through the tail and into the pillar to accept a brass taper pin.

Finally, test that the rack-tail rides over the snail as you advance the hands. If it does not, ensure that the filed angle is contacting the step of the snail, bending the rack-tail to suit. If it will still not ride over, reduce the thickness of the rack-tail on its rear face to weaken its 'spring'.

LOOSE FLY

Toolbox 2

When you inspect a striking clock, you need to examine the fly tension on its arbor. The fly has two jobs; it regulates the speed of strike by acting as an airbrake, and it reduces the locking force upon the gear train when striking comes to a stop. When the strike train locks, all of the force and momentum which carried that gear train in one direction is stopped abruptly by a detent, which deflects the force back through the gear train, causing it to bounce. This bouncing puts a huge amount of stress on the delicate wheels and pinions. However, with a fly which is able to slip on its arbor, as the train bounces back the fly continues to rotate forwards, with the friction with which it is fitted, dragging the arbor forwards with it. This reduces the amount of bounce and disperses some of its energy.

With the fly playing such an important role it is a surprise that it is so often overlooked by repairers, who occasionally leave it loose, completely nullifying its purpose, or even soldering it to the arbor, increasing the forces of locking and making the situation worse than in its natural state. Modern chiming clock flys are sometimes a solid fit, but the centrifugal action of that specific design reduces the negative force.

The fly should rotate with light finger pressure while holding the arbor still, but it should not spin freely. There are many designs of friction setting for a fly, all of which require a groove cut into the arbor to locate the friction spring. The majority are either a brass or steel leaf spring which presses against the arbor, either riveted as in English clocks, or held by friction as in French clocks.

To increase the friction of the fly is usually a case of removing the arbor and bending the friction spring to act against the arbor with more force. Sometimes these springs are so worn that they must be replaced. For a steel spring, a small section of watch mainspring is ideal; it can be cut and filed to shape easily. Alternatively, you could use a piece of brass sheet which has been hammered to increase its hardness, cut to shape, drilled and riveted to the fly with a brass rivet made from a piece of brass wire.

For the friction type with no spring, find the small dent on the outside of the centre tube of the fly which creates the 'pip' inside the tube and, using the staking set and an appropriate stake, tap lightly with the hammer to increase the depth of the pip.

Disassembled French fly.

Holding a piece of watch mainspring to file a new fly spring to size; remember to temper the steel before bending to shape.

Tightening the 'pip' in a friction-type fly.

SPLIT PIN HOLES

Toolbox 3

When a taper pin has been forced too deep into a cross-drilled hole, there is a likelihood that the stud or pillar through which the hole is drilled will split.

Many clocks have steel studs screwed into the front plate of the movement, behind the dial onto which the rack, lifter, minute wheel, etc., are pivoted. Because of their small diameter it is common for them to split. I often see split posts riveted over so that the minute wheel of longcase clocks cannot be removed.

When a stud is split and bent over, my favourite method of repair is to straighten the stud as close as possible to its original state and silver-solder to fill the tear and the cross-drilled hole. Provided it is done well, the tidy and strong solder joint can be filed and polished back to shape and the hole re-drilled. The only evidence of this repair is a slight discolouration towards the end of the stud where the solder does not match the steel work.

In the case where a rivet has been formed on the end of the stud, you often have to file away the end in order to release the component. A new stud will need to be made as a replacement and this is a turning and filing exercise to be done in the lathe.

Dial feet are often broken away at this point too. The common bodge for this is to drill through the side of the front plate and into the dial foot. A pin can then be inserted through the side of the plate to hold the dial foot in place. Often I find that pins secured this way come loose, the repair is ugly and it causes permanent damage to the movement. Depending on the dial type, there are some ways around it. The easiest to repair are the dial feet of silvered clocks, which can often be unscrewed from the dial to work on. If they are riveted in place you can work with the dial face down on a soft material to avoid scratching the surface.

Split taper pin holes are common where taper pins are driven too hard into their holes.

Use pliers to reshape the post as much as possible.

To repair a dial foot of the more common type, as seen in longcase clocks and bracket clocks, it is first sensible to familiarize yourself with the design. Can the dial foot be removed without damaging the dial? Is the dial foot riveted to a painted dial? Will manipulating the dial foot cause any damage to the dial paint or finish?

If the dial foot can be removed then it is best to do so. If not, proceed carefully and check the condition of the dial regularly as you progress. Cut or file away the broken portion of the dial foot down to the shoulder, being careful not to cut into the shoulder and reduce the effective length of the dial foot. Locate the centre of the stud and make a small mark using a centre punch to make a light indentation.

With the centre marked, select a drill which fits comfortably but is not tight in the hole which the dial foot fits. It is best at this stage to drill a test hole in a sheet of brass and compare its diameter to that in the plate. Drills often cut oversized holes, especially if they have been resharpened. With the drill selected, you need to drill out the centre of the dial foot to a depth of about halfway. If the drill is large, it is best to select a drill about a quarter of its diameter to create a pilot hole; this will help keep the drill on centre. Remember to protect the dial and use a drill press if possible. If you have not got access to a drill press, use an engineer's square as a visual guide for keeping the drill upright. Progress slowly and remember to clear the chips frequently by removing the drill from the hole: if the drill snags it may break the rivet and ruin the dial. If the dial foot was removed from the dial prior to drilling, the work is best done in the lathe, with the drill bit being advanced by the tailstock or in a pin chuck by hand.

With the dial foot drilled, you need to fill the hole with a piece of brass rod. Turn the brass rod down to fit the hole with a good interference fit and part it off long enough to fill the entire depth of the hole. Remember to leave enough length to extend through the plate by the same

Fill the post with silver solder to form the repair.

The set-up for re-drilling the cross-drill hole for the taper pin.

amount as the other dial feet. When you have a good interference fit at one end, and a comfortable fit without freedom in the plate, shape the end of the rod to match the design on the other feet.

To fit the rod to the dial foot, I like to use a drop of Stud-lock and a good interference fit between the components. I prefer not to solder as the amount of heat needed to warm the dial foot and the insert is enough to mark a dial. You could equally thread the dial foot hole and the insert, but it is unnecessarily complicated to do so and the result is no better, thanks to modern adhesives.

Smear a thin layer of Stud-lock on the rod and press it into the dial foot, with the dial firmly supported on a protective surface. When it will no longer press by hand, tap the end lightly with a nylon mallet to ensure you do not damage the shaped end of the stud. Light taps as if riveting should suffice.

The final job is to locate and drill the hole for the cross pin. Mount the dial back onto the front plate and pin it in place with the remaining dial feet. With the shoulder pressed hard against the movement, take a sharp knife and mark the new dial foot flush with the front plate. Remove the dial and select a drill by finding one to fit a hole in the other dial feet. The diameter of the drill defines the position of the hole, and you need the edge of the hole to be right on the line you marked. To position the hole, fit the drill bit to a flexi-shaft drill or hand-held rotary tool and lightly drill a dimple, by eye, in the correct position. With this approach you can lead the hole towards or away from the line until you have a dimple of the full diameter in the right position. To complete the hole you can either continue with the hand-held drill or return to the drill press, remembering to site the drill to pass through the centre.

With the hole drilled, I like to finish with a cutting broach to better grip the tapered pins. Try the broach in a hole of an existing dial foot and mark the depth to which it enters. Broach open your drilled hole to match this depth; this way you can ensure that the same size pin will fit all holes and save time during reassembly of the clock for future repairers. Chamfer each end of the hole to remove burrs.

Always broach taper pin holes so that their taper matches the pins; this ensures a better grip which will not come loose.

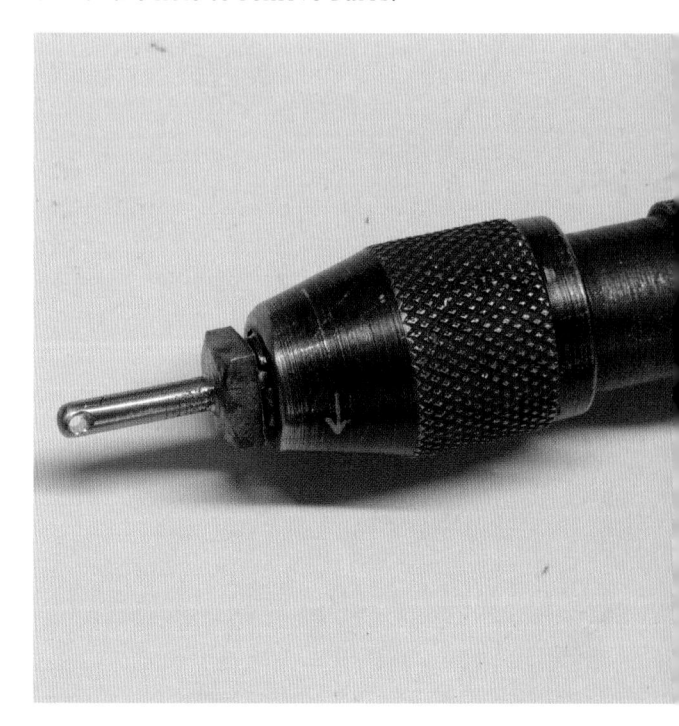

Finished post repair, now filed and polished to shape.

LOOSE PILLARS

Toolbox 1

Loose pillars come in many forms. Screws and nuts can be easily tightened but riveted pillars must be attended to properly.

Over time clocks are taken apart and put back together repeatedly and the rivets can work loose. Any loose pillar is a potential problem to the running of the clock so it is the repairer's job to remake the rivet between pillar and plate.

When originally riveted, the pillar would have had an excess of material with which to make the rivet. This would have then been filed flush with the surface of the plate. This resulted in a near-invisible rivet, which unfortunately you are now unable to replicate. Without the excess of material to make the rivet, you have to blemish the remaining rivet in order to remake the connection.

To minimize damage to the plate and pillar, use a domed punch from the staking set. Rest the pillars face down against the firm surface of the workbench. Work the doming punch around the edge of the original rivet whilst gently tapping with a hammer, using enough force to subtly distort the rivet without leaving deep marks. Repeat this process around the edge of the rivet until the connection is solid and does not move when tried by hand.

Finish by gently filing to remove high spots, followed by graining or polishing to the desired level of finish. It is not necessary to try to remove every mark and doing so would be detrimental to the life of the clock as future repair attempts will be hindered by the lack of material left in place.

TURNING

Whole books have been written on the use of lathes for both hobby and industrial use, so it is hardly possible for me to cover the subject in depth in this brief section. However, I have not read a single one of these books: I learnt to use the watchmaker's lathe with a hand graver through experimentation and advice from experienced watchmakers and turners. These tips and experience transferred over to the larger 10mm Pultra, which I mainly use for the cross slide and drilling tailstock. Provided you understand a few basics, the best work is achieved through experience and patience.

All lathes work on the same basic principles, although they are different in layout.

- Keep your cutting tools sharp. A blunt cutting tool requires extra pressure to force the cut, which nearly always results in a broken or bent work piece or a poor surface finish. Keep a fine diamond lap nearby and periodically sharpen your graver; use a honing guide if necessary to maintain the cutting angle.
- The finish of the tool is reflected in the finish of the work piece. The surface of the cutting edge is imparted directly on to the work piece as it cuts. When there is a lot of material to remove during sharpening (to remove a chipped edge), use a coarse diamond lap or stone, but always finish on the finest grade available to achieve a polish.
- Cut on or slightly above the centre line of the work piece, never below. When you cut below the centre line of the work piece, the graver is liable to catch and be dragged under, forcing the work piece upwards, snapping or bending it. Ideally you cut exactly on the centre line. When using a cross slide, with the tool gripped firmly in the holder, you set the height to match a dead centre mounted in the headstock. However, with a hand graver this is difficult to maintain. I find it much more reliable to cut just a fraction above the centre line, creating a buffer for any errors of judgement which would otherwise allow the graver to pass below centre.
- Keep cutting speeds low to moderate. Lathes were hand-driven when these clocks were being manufactured. I prefer to use relatively low cutting speeds of around 300 to 500rpm (not a significantly important figure) and my reasons are as follows: chips ejected from the work piece hurt a lot less when travelling at lower speeds; you are often hand-turning small pieces and working with your face very close to the headstock so lower speeds are much safer; and it keeps the noise down.
- Collets are more accurate than three- and four-jaw chucks, so use them. Collets clamp

the work piece around its entire diameter, holding it securely and concentrically to the headstock spindle. Three- and four-jaw chucks clamp tightly only at three or four points. They can potentially break the surface or raise burrs if the jaws are loose and slip during use, or are so tight that they dig in. Because of this, collets are able to hold fine hollow components which would be crushed by a jawed chuck. Most watch- and clockmaker's lathes are designed primarily to use collets as their method of holding the work piece; adaptors to fit various chucks can be of use for holding large components which the collets will not accept, but should otherwise be avoided.

- Support the work piece thoroughly. If the work piece is short and thick, then the support of the collet alone will be enough for turning, but as pieces get longer, they will deflect as the cutting tool is applied, causing inaccuracies and possibly breakage. Long thin work pieces should be supported at each end by both the head- and tailstocks provided; if necessary, use a three-point fixed steady or travelling steady to support the work piece near where the cut is to be made.
- Your cutting tool must be harder than the material of the work piece. Carbon steel gravers are hardened and tempered just enough to relieve the stresses in the material. They should not be used to cut any material harder than blued steel. If you are turning a harder material, a tungsten carbide graver should be used; these are more brittle and prone to chipping at the tip.
- Work in good light. Good light is essential for good work – this one speaks for itself.
- Get comfortable. You may be sitting or standing at the lathe for several hours at a time making delicate components, so it is important that you are comfortable. Raise the height of the lathe with a custom-made table so that you are not sitting hunched, and position the speed control where it is easy to reach.
- Consider your safety before starting any job. Before starting work at the lathe, tuck in any loose hanging clothing, tie back long hair and beards, and wear some form of protective eyewear (wearing a loupe will protect the eye).

SOLDERING

Soft (or electrical) solder is to be used sparingly, and not in situations where structural strength is important. For example, a good use of soft solder would be to stop a component from slipping, such as the connection of hammer rod to hub on French movements, or a large bush.

Silver soldering produces a much stronger joint and is similar in use to brazing. It is used in situations where a component has broken. A common example would be where a longcase pendulum rod has snapped off at the bob: repairing by re-threading would shorten the rod by about 1.2cm (½in), while replacing the rod would ruin the originality of the clock.

Good soldering is all about cleanliness. Use emery paper to clean all surfaces back to bare metal in the area of the repair. If you are soldering a crack, the only way to be sure it is thoroughly clean is to trace it with a very fine piercing saw, following the crack as perfectly as you can.

For flux, borax mixed into a paste with water was traditionally used, but there are now commercial fluxes which surpass the qualities of borax and are easier to use. Apply a small amount of flux to the area to be soldered, enough to cover the area and allow capillary action to draw it into any cracks.

The role of flux is to form a barrier between the oxygen in the air and the clean metal. Without it, the material will oxidize before the solder reaches melting point, and the oxidation will prevent the solder from flowing.

Heat the entire work piece if using a small blowtorch, otherwise the heat will be dissipated away from it is needed. Larger components with more material act as a much stronger heat sink than smaller ones, so if you are soldering thin components you may be able to skip this. If you are soldering a small area on a large component, it is best to use the more powerful flame of an oxy-propane torch which will work fast enough that the heat will not be drawn away.

Soldering small components with soft solder can often be done with a soldering iron. The standard rules apply here also, but remember to keep the tip of the iron clean and 'tinned' with solder after use to increase its life.

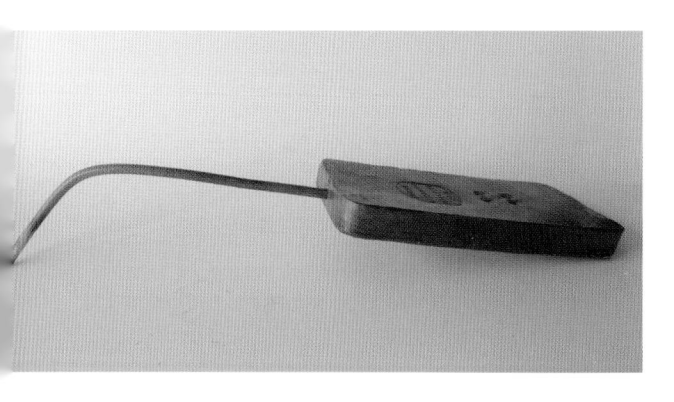

Homemade soldering weight used regularly for holding parts in place during soldering.

I have made some useful soldering weights to keep components still while soldering. It is important that the weights do not act as a heat sink, so the design is critical. Mine are simple lead bars with a protruding wire at one end. The wire is shaped in a manner that will help grip the parts to be soldered, and is long enough that the lead will not be heated.

You should set up a separate area for soldering; use soldering blocks to absorb the heat and reduce the risk of fire. You will also be using these soldering blocks for hardening, tempering and bluing. You will need good light to spot colour changes in the materials.

RIVETING

Riveting is a technique used in many clock repairs and must be understood and mastered. It is possible to get a very good mechanical hold between components using a riveted joint. It is commonly used for dial feet, pillars, rack-tails, large bushes, wheels and pinions, pallets, balance staffs, hand collets, suspension repairs, barrel hooks and so on.

Riveting spreads metal to produce a clamping effect between two components. For example, to join two pieces of sheet material together, clamp them together and drill a hole through both pieces. Fit a taper pin into the hole until it is tight. Cut off each end of the taper pin, being sure to leave a small amount of material protruding. With one side rested on an anvil, lightly tap the rivet repeatedly with the ball end of a ball-pein hammer or with a domed punch where accuracy is necessary. Work your way around the rivet until the protruding material becomes domed like the head of a mushroom. Flip the work over and rivet the other side. If you had chamfered the hole first, you could now file the rivet flush, and the riveted material will have spread to fill the chamfer. Without the chamfer, leave the domes of the rivet in place to form the two clamping heads which will keep the materials together.

When you rivet during the repair of clocks, you often do not use a separate riveting material, instead using part of one of the pieces to be joined to form the rivet. Examples would be riveting the rack-tail in place using the rack post to form the rivet, or riveting over a pinion or collet to mount a wheel.

AUXILIARY REPAIRS

HANDS

Toolbox 3

Every time the clocks change, I get at least one customer coming into the shop with a guilty look on their face and a clock with a broken minute hand. Proper repair takes between fifteen minutes and an hour, depending on the size and finish of the hand in question.

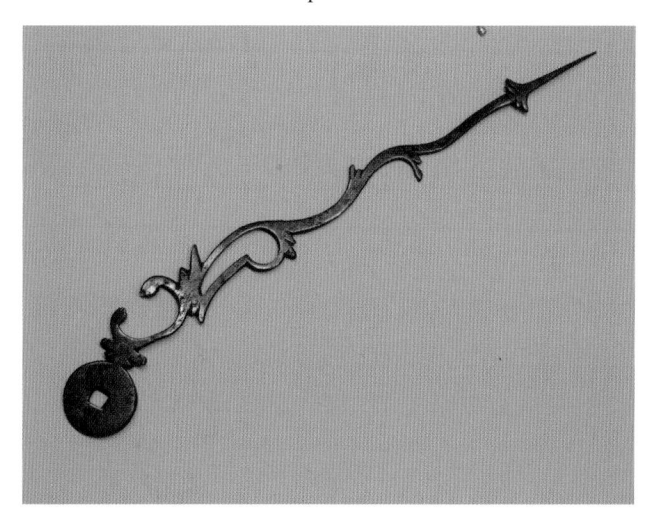

A common breaking point for longcase minute hands can be seen near the base.

Steel hands should be silver-soldered back together, which would ruin the finish of a gilded hand. Small carriage clock hands would be obliterated by silver soldering, but as they are under glass and never touched, they can be soft-soldered. Hands which have been broken into several parts should be replaced. Common sense should be used when determining whether or not to replace the hand entirely: originality is an important factor in restoring antiques. A well-repaired original hand is preferable over a modern replacement in some cases; the only other option is to make a replica of the original.

To repair a blued steel hand, first thoroughly clean the area around the joint, front, back and around the edges using a emery paper or a fine file, then clean the break itself with a fibreglass scratch brush. On a fire-resistant brick, place the two components face down and so that they line up, then pin them down with soldering weights and apply some flux. Heat the hands until the flux begins to flow. Place a small piece of silver solder across the joint. Reheat the hands and the solder should flow into the joint and across the back of the hand. Allow it to form a small build-up of solder across the rear of the break: this will add strength to the repair and be invisible from the front. If you find that the solder will not flow, add a little more borax and try again. If it still fails to flow, stop now before you ruin the hand by overheating; let it cool down and start again, re-cleaning back to bare metal.

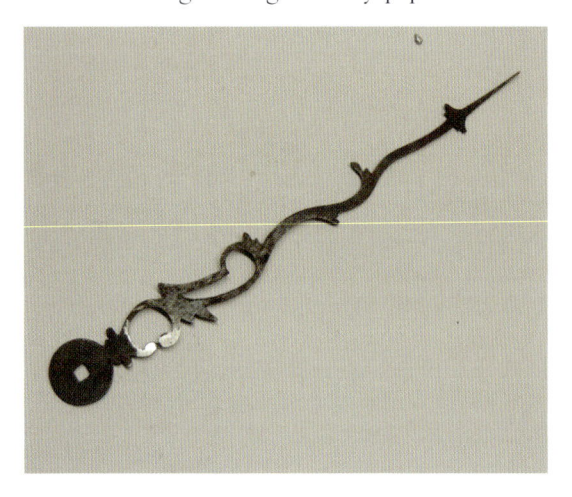

Clean the steel back to bare metal around where the repair is to take place; any dirt or oxidation will hinder the repair.

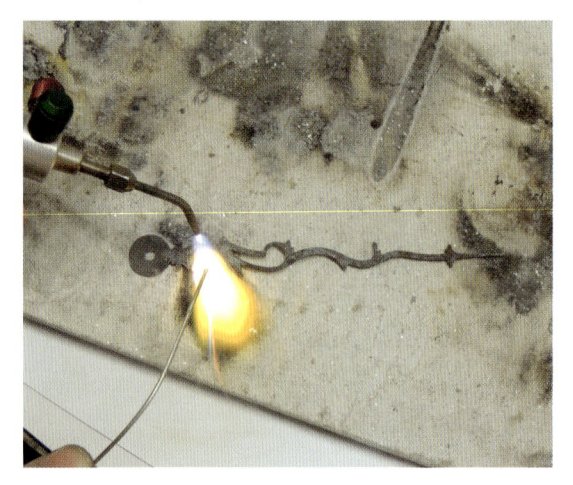

Apply flux and silver solder to the rear of the repair area.

Once the solder has flowed evenly across the joint, leave the hand to cool naturally.

When the hand is cooled you will need to remove the oxidation and remains of flux. A pickling solution or acid is good for this although you could hand-clean back to bare metal as before. Take a file and remove the excess solder; if you have used the right amount this should not be necessary. Clean the entire hand with emery paper of increasing grades until you achieve the desired finish. Do not touch the hand with your fingers; the oils on your skin will ruin the next step.

Blue the hand by heating evenly with a gentle flame from the blowtorch. If you go beyond the desired dark blue colour, re-clean the hand and start again. I keep an anvil nearby to act as a heat sink; if the colour is changing rapidly, place it over that part to slow the process.

Once the hand is blued you will notice a small sliver of silver solder visible from the front of the hand along the break. This is a good sign that the solder penetrated and is effective. Because solder does not blue with heat, you will need to colour it to disguise the repair. Take a blue permanent marker pen (Sharpie is one relevant trade name) and colour the solder; before it dries completely, give it a light rub with a finger to remove the telltale metallic shine. Finally, wipe the hand over with an oily rag: this will darken the blued finish and provide rust protection to the hand. The soldered joint should now be unnoticeable to the untrained eye.

To repair a small blued carriage clock hand is a job for a soldering iron and soft solder. If

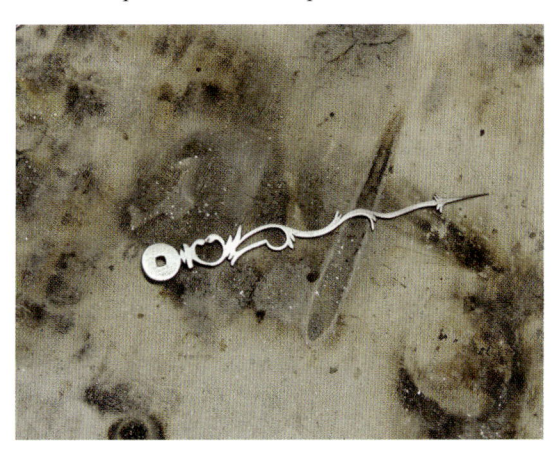

Clean the entire hand by graining lightly with emery paper or polishing.

Bluing the hand with heat; keep the flame moving and do not let it rest for too long in one area.

On the finished hand, colour the silver solder at the joint with a blue marker pen and it will barely show; leave a mound of solder on the rear to strengthen the joint.

A broken carriage clock hand is not uncommon and is usually the result of fouling hands or well-meaning amateur repairers.

you can find an adequate replacement, I would recommend fitting this instead.

Clean the hand around the breakage with a glass scratch brush, but only at the rear of the hand. Apply some soldering flux around the joint and melt some solder onto the end of the iron. Whilst pinning the two parts of the hand with the tips of your tweezers, rest the iron over the joint. As the hand warms, the solder will flow and coat the back face. I do this face down on a sheet of glass to provide a perfectly flat surface. This is not a strong joint! However, as the hand is behind glass it is enough. Fit the hand to the clock and ensure that the solder does not foul the dial or other hands; if it does, scrape some solder away with a sharp blade.

You cannot blue a soft-soldered hand with heat: the solder will soften first and ruin the joint. Instead, if necessary, use a blue permanent marker pen to colour the hand as evenly as possible. You may need to colour the other hand to match. This repair will last a very long time provided the hands are fitted with care and not touched by the owner.

Repairing gilded hands requires an extra step. Silver-soldering the hand as you did with the blued steel hand will ruin the finish. Instead, polish or grain the hand to resemble the original finish and thoroughly degrease both the minute and hour hands. They will both need to be refinished so that they match in colour and texture. Apply gold leaf to both hands following the manufacturer's instructions. Alternatively, send both hands off for electroplating.

When the hand is broken in several places it is best to replace it with a hand-made replica. This is a hand-piercing exercise which requires patience and practice. Templates can be found online and printed to scale. This template can then be glued to a piece of steel and hand-pierced.

Clean the joint area with emery paper, as with all soldering jobs.

Apply flux and soft solder using a soldering iron; pin the parts to the table as shown here to stop them moving while bridging the gap with solder.

The solder bridge does not look pretty but can be reduced with a file to improve the situation.

From the front the repair is not noticeable and the blue of the hands has not been ruined.

BELLS

Toolbox 3

Cast bells are very brittle and crack easily. You will instantly recognize a cracked bell because it rings with a dull thud. Bells are easily repaired and should not be replaced. In the picture, the red line follows a typical crack.

There are several methods for repairing a cracked bell but they always involve finding the crack, so you must start with crack detection. Finding hairline cracks can be difficult, but I have never failed to spot one by eye. Take a loupe and study the outside rim of the bell as you turn it in your hands. When you think you have spotted a crack, check the inside edge of the bell. The crack will go right through; if it does not then it will not interrupt the vibrations or the ringing of the bell and is not what you are looking for. If you are struggling to spot it, you can buy crack detection inks which wipe off and leave a trace of bright red ink in the crack.

Having located the crack you should take a minute to decide which repair method to use. Hairline cracks solder nicely and leave you with a very clean-looking repair, while larger cracks can actually be opened up to improve the situation.

If the crack is a wide one you need to locate the point at which the crack ends. Starting at the outer edge of the bell, follow the crack in towards the centre and place a small ink dot where the crack finishes (as indicated by the white circle in the picture); do not punch a centre, as this will crack the bell further. Drill at this point: this stops the crack from propagating in the future. Take a sharp twist drill of an appropriate diameter. (You need to remove the terminal point of the crack, and a larger diameter drill provides a buffer against the unseen. Use 5mm as a starting point for a longcase bell.) Using a pillar drill, carefully drill through the bell. Do not push hard or go too fast. If it will not take, file a small flat where you intend to drill, just enough to stop the drill wandering.

With the hole made, you can open up the crack: this is to allow the bell to vibrate with freedom. (The dull thud of a cracked bell is due to the two faces of the crack vibrating against each other and 'cancelling out' the harmonic vibrations.) You will need a piercing saw, a

A cracked bell will not show up in photographs; drill at the white dot to remove the root of the crack, although soft soldering is the preferred repair if possible.

loupe and a fine marker pen. Whilst wearing the loupe to ensure accuracy, use the marker pen to trace the crack; this makes it a lot easier to follow as you are now going to take the piercing saw with an appropriate blade and chase the crack all the way up to the drilled hole.

When the job is complete, file away any burrs and the bell will ring as if it were new.

The preferred method of restoration for a cracked bell is soft solder. For this you will need to use a micro-torch; the soldering iron will not be hot enough. Thoroughly clean the bell before you start; I always use the ultrasonic tank for this, with a brand-new rinse solution to ensure that no grease or grime remains in the crack. Dry it thoroughly with a hairdryer and scratch brush around the crack on both the inside and outside, and do the same along the rim. Apply flux to the inside edge of the crack; capillary action will draw it into the crack itself. If you are using a flux paste, it will flow into the crack as heat is applied. Using the micro-torch, heat the entire bell; this will stop heat being wicked away from the crack. When the flux starts to flow, apply soft solder to the full length of the crack. When you can see solder covering the entire crack, run the flame over it once or twice just to ensure it has flowed through, and leave it to cool. Thoroughly clean off any flux or excess solder which appears on the outside of the crack. Wire-brush, polish or soak the bell in a gentle pickling solution to remove any oxidation.

PULLEYS

Toolbox 2

Pulley wheels break in various ways. They are the link between the large driving weight and the movement so are subjected to a lot of force, especially during winding as the weights are prone to swinging and bouncing if not supported by hand. The pulley wheels turn constantly on their arbors, albeit very slowly as the clock runs down, and then much faster when wound, so they are subject to wear in the same manner as any pivot. They are often overlooked during overhaul by many repairers, but they should be cleaned and oiled along with every other mechanically functional component of the clock.

The most common failure is a worn centre hole of the pulley wheel, and this is what you will repair here. To begin you will need to disassemble the pulley wheel. There are generally two types of pulley, those with a riveted arbor

Use hand removers to break the rivet at the centre of the pulley assembly.

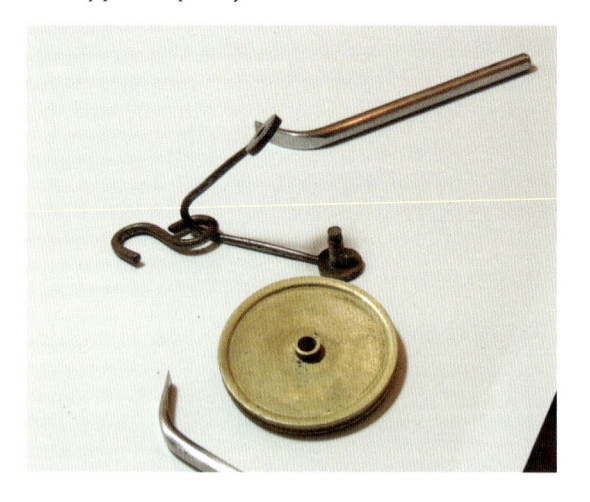

Fully disassembled pulley wheel, ready for repair.

Broach open the centre hole of the pulley wheel to accept a large bush.

and those with a long shoulder screw. We will focus mainly on those with a riveted arbor, as the repair transfers across both types.

To remove the riveted arbor I use a sturdy pair of hand removers to force the strap away from the wheel. This will cause the rivet to deform in order to pass through the hole, which allows you to close the rivet again upon completion of the repair. Drilling or filing to remove the rivet would require turning a new arbor in order to reassemble the pulley.

To repair a worn centre hole is a bushing exercise on a large scale. You could open the centre hole to accept the bush on the lathe, using a large broach or with the pillar drill. The bush will be made from brass tube or bushing wire.

Use bushing wire or brass rod to ensure a full width insert; clock bushes are not tall enough to bush most pulley wheels.

Generously apply soft solder flux to the insert before placing it in the broached hole.

Soft-soldering a large component like this will require the heat of a blowtorch rather than the soldering iron.

Open the centre hole of the wheel to accept the brass bushing wire; cut the bush to be slightly oversized in length and cover its outer edge with soft solder flux. Press it into the centre of the pulley wheel and heat the wheel with a blowtorch. Apply a small amount of soft solder to the raised edge of the bush and it will be drawn into the joint.

Alternatively, you could chamfer the edge of the centre wheel hole after broaching it to accept the brass bushing wire or rod, and rivet it in place from both sides.

Select a drill bit the same diameter as the arbor and use the pillar drill or lathe to open the centre hole, following this by broaching until it is large enough to spin freely on the arbor. Use the lathe or hand files to remove the excess length of the bushing wire and finish by graining. Reassemble the pulley wheel and remake the rivet on the arbor.

Clean up the wheel for reassembly by turning in the lathe; broach the hole to be a good fit on the arbor.

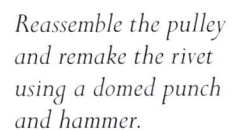

Reassemble the pulley and remake the rivet using a domed punch and hammer.

The fully repaired pulley should not show any signs of repair.

PLATFORM REPAIRS

JEWEL-HOLES
Toolbox 3

Cracked jewel-holes are common and you will come across them sooner or later. A cracked jewel-hole should not be left in place; it will cause excess friction and wear on the arbor or staff. Replacing them is an intermediate technique and, due to the limitations in tooling and spare parts you are likely to face, it may be best to outsource this type of work to a professional.

There are two main methods with which jewel-holes are fitted to platform escapements: press-fit jewelling used in modern platforms, and rubbed-in jewelling used in antique platform escapements. Special tooling is required for both of these jobs: a jewelling press, and a set of rubbing tools specially designed for rubbed-in or burnished jewel-holes.

So how do you tell what type of jewelling you have? By appearance. Rubbed-in jewels have a unique design in which a hole is bored into the plate for the jewel-holes to be set in. A thin ring of brass is then quite literally rubbed or burnished over the jewel to form a cage to hold it in place. Press-fit jewels are ground to a very specific outside diameter, and are pressed into an accurately cut hole in the brass plate; the natural elasticity of the brass allows it to stretch and grip the jewel to hold it firmly in place without chipping. Rubbed-in jewels cannot be pressed into the plate; they are not perfectly round and will break when pressed into a round hole.

Fitting any type of jewel can be a quick and easy task with practice, provided you have a selection of jewel-holes to hand. For rubbed-in jewels, I keep a selection of scrap platforms; using a sharp graver I am able to cut free the jewel-hole which best fits the pivot. For press-fit jewels I have an assortment which I bought from my materials supplier.

Rubbed-In Jewels

When faced with cracked jewel-holes, first find a suitable replacement, and I will assume that you have an assortment and have selected the jewel which best fits the pivot.

The old cracked jewel can be pressed out of the setting with the sharp end of a strong pair of tweezers. It will smash in the process but this is no loss. With the setting cleared, you can insert the opening tool and, with a small drop of oil to prevent tearing the brass setting, tighten the screw to spread the prongs of the tool; increase the pressure slowly as you rotate the tool to open the setting. Open the setting only enough to drop the new jewel-holes into place; any extra creates a higher risk of cracking the thin work-hardened brass setting.

Remove the old cracked jewel by pressing it out with a sharp punch or tweezers.

Use the opening rubber to spread the setting; apply a drop of oil to prevent tearing.

The new jewel hold should sit neatly within the setting.

Use the closing rubber to rub the setting down onto the new jewel hole, trapping it in place.

Completed rubbed-in jewel hole repair.

With the jewel-holes fitted in the setting, use the closing tool, with a small drop of oil again, to burnish the setting back over. There is a risk of cracking the jewel-holes or the setting if you rub it down too hard, so when the jewel is secured, stop.

Finally, run the components through the cleaning station to remove the remaining traces of oil. When everything is dry, apply a small drop of Stud-lock to the edge of the setting using a watch oiler. Capillary action will draw it in around the jewel-holes and will hold it firmly in place. In this situation, Stud-lock reduces the risk of damage to the jewel and the setting; it is not visible and is perfectly reversible for future repairs.

Press-Fit Jewels

These are in common use today and can, in some situations, be used as a replacement for rubbed-in jewels where originality is not a matter of great importance. If this is the case, broach open the setting and fit a bush, riveted from both sides to ensure it holds as the jewel is pressed in. If you can find a jewel with a larger outside diameter than the original hole, there is no need to bush the plate first.

A jewelling press is needed for this job, with a full selection of reamers and pressing tools. Set up the jewelling press with the reamer which just enters the hole to be broached. By hand, rotate the cutting tool to broach open the hole, working your way up the sizes of reamer until the jewel-hole half-enters the hole. Swap the reamer for a flat-faced pressing tool and fit the lever handle to the jewelling tool. Remove the jewel from the hole and set the depth stop of the tool so that the face of the pressing tool is level with the top surface of the plate when fully pressed home. Fit the jewel-hole in place and, using gentle and constant pressure on the lever, press it all the way down until the depth stop is reached. The result will be a jewel-hole which sits level with the plate. If you need to increase the end-shake of the arbor adjust the micrometer end stop by the appropriate amount and press the jewel in a little further.

Use the jewelling press with its micrometer depth stop to press a jewel into its hole.

BUSHING PLATFORMS

Toolbox 3

Not all platform escapements are fully jewelled. Many have jewelled balance pivots, but the lever and escape wheel pivots remain in brass which can wear quite badly at times.

To bush pivot holes on this scale, you use the same techniques as bushing at a larger scale, but the tools are much smaller. However, at this scale any mistakes in uprightness can be much more troublesome than you are used to. When bushing platform escapements, I finish the job in the jewelling press to ensure absolute uprightness of the bush, and to provide a means to press it into place under control, rather than driving it in with a hammer and punch, which could distort the plate.

You still lead the hole over as you did with a file when bushing larger plates, but at this scale you use a small broach. Insert the broach into the hole, but not so far as to bind, and push it hard against the side of the hole from which you intend to remove material. To prevent the broach from tipping over, apply sideways pressure from each side of the plate.

With the hole lead over by the appropriate amount, broach it open until it is round. Move to the jewelling press and follow the procedure for fitting a press-fit jewel, only this time using a small brass bush in place of the jewel-hole. When the bush is fitted you will need to broach the hole open to fit the pivot. Do this with the jewelling tool and its selection of reamers if possible. If the hole is too small, use hand broaches to open the hole and aim to finish with the jewelling tools' smallest reamer if possible.

HAIRSPRINGS

Toolbox 1

Hairsprings take a lot of practice and technical understanding to manipulate for good results. There are, however, a few small jobs that any beginner should be able to complete with a bit of practice and some basic knowledge.

Setting for Beat

When a platform escapement is out of beat, you set it right by rotating the balance spring on the staff. If you recall from our discussions on the theory of platform escapements, the hairspring is mounted to a split collet, which is a friction fit onto the balance staff or cylinder. This all allows for adjustment when setting up the clock.

Remove the balance from the escapement before starting the work to prevent damage to the pivots.

Cylinder escapements are the easier to set for

beat as there is a little trick you can apply. With a pair of tweezers, grip the cylinder across the impulse faces as in the picture. The stud for the hairspring should line up with tweezers; if not, adjust the spring until it does.

To make the adjustment all you need is a spare screwdriver. File the blade to a shallower angle than you would usually use for removing a screw; this reduces the risk of slippage in my experience. Insert the blade into the split of the collet, wherever there is adequate clearance. Twist the screwdriver to loosen the collet's grip of the cylinder, and twist the collet into its new position. One slip and the hairspring could be ruined, so work carefully or outsource it.

For a lever escapement there is still a rule you can follow, but unfortunately you will have to reassemble the platform after each adjustment to check it. The impulse jewel should hold the lever in the centre of the banking pins when there is no power on the escapement. An out-of-beat lever escapement will set the lever to one side.

To adjust, the procedure is the same as for a cylinder escapement. Remove the balance from the balance cock and adjust the hairspring collet by rotating it with your spare screwdriver. Remount the balance in the platform and check the resting point of the lever. Repeat this process until you have got it right.

On modern lever platform escapements, the mounting point for the balance spring stud is sometimes adjustable, so it is not necessary to adjust the hairspring. In fact, you can adjust the beat while the clock is running with no ill effects.

Regulation

Platform escapements are regulated by adjusting the length of the hairspring, which is a simple case of moving the index left or right as labelled.

However, in many cases it is possible to regulate the clock by adjusting the hairspring to lean against one of the curb pins; this keeps it in constant contact, effectively shortening the spring. In normal use the hairspring is able to bounce between both pins, only being in contact with the pins for part of the escapement cycle. During the time it is free between pins, the hairspring is full length, or free sprung, making the clock run slow.

In a clock which is very slow towards the end of the week, it is common that the balance amplitude has dropped as the power available decreases. You may find that as the balance amplitude drops, the hairspring looses contact with the index entirely, the result of which is slow running. Close the index pins by squeezing them together with tweezers to reduce the freedom of the hairspring, but not so tight as to grip it.

Hairspring manipulation is an advanced topic which we cannot cover in more detail here.

CYLINDER WEAR

Toolbox 1

Cylinders wear tremendously if not regularly cleaned and oiled. The teeth of the escape wheel are in near-constant contact with the wall of the cylinder so a deep groove is cut into the wall over time, and the impulse faces are worn away.

This wear causes two major problems: it weakens the cylinder wall dramatically and it effectively moves the impulse faces of the cylinder away from the escape wheel.

The ideal repair here is to replace the cylinder, but this is an advanced job which is best outsourced to a specialist. However, as many cylinder clocks do not justify the cost of a new cylinder, you can often find a workaround. Fitting a new platform is an option; however, this

Gripping a cylinder across the impulse faces like this reveals the in-beat position for the hairspring stud.

is rarely more cost-effective than replacing the cylinder.

If you are repairing on a shoestring budget then the best repair is often to simply move the escape wheel to a position where it is no longer running in this groove. First check for clearance in the cylinder. The following method only works with horizontally mounted platforms in which gravity keeps the arbor against the lower cock.

Wearing a loupe, peer into the side of the escapement to view the clearance around the escape wheel when a tooth is within the cylinder. If the clearance below the arm of the tooth is enough that the escape wheel will not clash with the cylinder, then it is safe to proceed.

To lower the escape wheel by the appropriate amount, strip the platform to avoid damage to any of the components. Place a slip of paper between the lower escape wheel cock and the main plate and reassemble. This will drop the cock by the thickness of the paper, approximately 0.1mm for standard 80gsm office paper. You should notice that the escape tooth now rests just below the wear groove on the cylinder. Trim the paper so that the repair is not visible. This is a temporary repair which only delays the inevitable replacement of the cylinder until the next service.

The end-shake of the arbor will be slightly increased but not so much that the pivot does not locate in the upper jewel-holes.

PALLET WEAR

Toolbox 2

Often the drops of the recoil escapement are uneven; the result is a clock with a very annoying tick and a lack of power. This has happened because, over the years, wear has been filed and polished out of the pallet faces, and the pallets have been closed, lowered, bent, etc.

To correct worn pallet faces or uneven drops you need to create a new pallet face on the anchor. This is done by soft-soldering a strip of mainspring onto the face and filing it to suit.

If the drops are even but the pallet faces are heavily cut then you do not want to modify the position of the pallet face. To add material you will first need to create a space for it by filing away the old pallet face by the same amount as the piece of mainspring you will be fitting. To remove the right amount of material, start by taking measurements from a datum point on the anchor as a reference. File the pallet face with a

A severely worn pallet face which will need to be addressed.

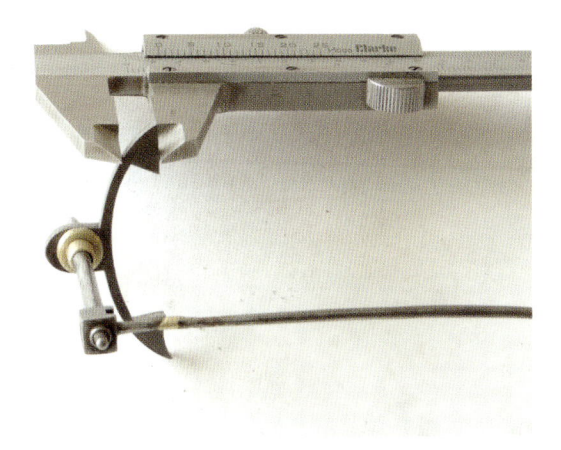

Measuring the thickness of the pallet face from a repeatable datum point.

Use a diamond file to reduce the hardened steel of the pallet face by the required amount.

Reduced pallet face now shows no signs of wear on its surface.

A slip of old mainspring cut to fit slightly oversized on the pallet.

With both surfaces tinned with soft solder, offer them together and apply heat to join them.

Remove the excess material of the mainspring by filing along its length; never file across it, which would break the solder joint.

With a bit of polishing the repair should be imperceptible to the untrained eye.

diamond file, or temper the anchor to light blue and file with steel files, being careful to maintain the curve and angle of the original pallet.

Break away a section of mainspring just oversized for the pallet face. The spring steel is hardened carbon steel and will snap easily if bent through 90 degrees.

Clean and flux both the pallet face and mainspring. Tin the mainspring but do not use the soldering iron so hot that it overheats and tempers the metal. We heat the anchor with the blowtorch, because the soldering iron will not be hot enough to raise its temperature sufficiently to melt the solder. Lay the mainspring in place on the anchor and gently heat until the solder flows. Press down with tweezers to push out any excess solder and hold this position until it has hardened

File away the excess material using a diamond file, working along its length to avoid breaking the solder joint. Polish the working face to reduce friction.

If the drops are uneven, reface the pallet with the largest drop without reducing the original face beforehand. Make final adjustments to even out the drops by filing and polishing the pallet faces.

BROCOT PALLETS

Toolbox 1

The pallets of the brocot escapement are so often incorrectly adjusted that you will almost certainly come across this problem early in your career.

The depth of the pallets into the escape wheel is adjustable at two points on the visible escapements; using the eccentric will raise and lower the anchor but will, at the same time, move it ever so slightly to the left or right, affecting the action of the escapement.

The front cock of the anchor can be lightly tapped up or down, raising or lowering the pallets without moving them sideways; this can be taken advantage of if the teeth are just catching on the pallets.

The pallet stones can be rotated in the anchor, and this is most commonly where the problems occur. They are held in place with shellac, although steel pallets may be a press fit. To adjust shellacked pallets, you will need to soften the shellac with heat. My preferred method is to remove the pallet arbor from the clock and support it in a vice by gripping the arbor. With a warm soldering iron, approach the pallet to be adjusted, but do not touch it. The heat radiating from the soldering iron will be enough to soften the shellac. In your other hand, use tweezers to put light pressure on the pallet; it will begin to move as the shellac warms. Adjust the distance of the soldering iron to keep the consistency of the shellac stiff but pliable as you make your adjustments.

Be certain that the pallets are upright in the anchor and adjust their position until the highest point of the curve becomes the landing point of the escape tooth.

As you rotate one pallet, bear in mind that you are not just rotating the landing point on that pallet, but the drop point also, which will affect the landing point of the other pallet. With a bit of trial and error, you will be able to correctly adjust the escapement.

CRUTCH ROD

Toolbox 3

The crutch rod is subject to constant bending every time the clock is moved or cleaned and subsequently set back in beat. Bending a piece of steel back and forth ultimately will lead to its failure, and in the case of steel crutch rods, this is most commonly in the form of delamination. Delamination is the separation of the layers within the material's structure and the result is a significantly weakened crutch rod.

Crutches come in several designs as detailed earlier in this book, and are made of either brass or steel. It is most commonly steel crutches on longcase clocks which become damaged and need repair, so it is the repair of these crutches which we will focus on here.

The crutch rod of a longcase clock is a thin steel rod of roughly 1.5mm to 2.5mm diameter; it is riveted at its bottom end to the steel crutch foot, and at its top end to the pallet arbor. The top rivet is prone to coming loose, and I have seen many poor attempts to soft-solder it back in place. The lower rivet tends to remain tight, but if the pendulum is knocked off during winding, then this joint is subjected to high forces when the pendulum block gets caught in the crutch foot as it falls, resulting in the crutch rod tearing near the foot. The final common failure is the crutch rod breaking somewhere near its centre, where it is bent for beat setting.

If the top rivet has previously been soft-soldered, this joint is likely to become loose during beat setting. To repair this once and for all, you will need to remove the crutch rod from the pallet arbor and remove all traces of solder. Designs vary from maker to maker, but in most cases, the crutch rod can be driven out of the arbor by a blow from above with a punch on the centre of the rivet. Do this with the arbor well supported on an anvil.

With the crutch rod removed, heat and scrub away all of the solder using a blowtorch and a brass bristle brush. The remains of solder on the surface will need to be filed away. You will often find that previous damage means that the parts are a sloppy fit and will not hold well if riveted as they would have been originally. If you think you can remake the rivet without losing stability of the crutch rod, then go ahead and assemble the parts, riveting the top of the crutch rod over to attach it to the arbor.

If riveting is not an option then you should attach the crutch rod by silver-soldering; this is strong and if done well can be very subtle. Instructions can be found in the section on soldering.

When the crutch rod is damaged near the foot it is often possible to remake the rivet by slightly shortening the rod. Hang the pendulum in place to check for clearance, if there is adequate clearance to raise the crutch rod by 2–3mm without missing the suspension block, then continue with the repair, otherwise repair by silver-soldering as before.

Break away what remains of the connection between the crutch rod and foot until the crutch foot is entirely separate; rest it upside down on an anvil and punch out the remains of the crutch rod using a hammer and punch. File the end of the crutch rod until the damage is removed, and then proceed to file onto it a shoulder and a reduced diameter which will pass through the hole in the foot and protrude enough to form a rivet. With the crutch rod clamped tightly in a vice, and with the newly formed pivot and shoulder facing up, place the foot into position and make the rivet with a ball-pein hammer. The shoulder and rivet will work together to clamp the crutch foot in place.

When the crutch rod is broken or damaged at its centre it is best to replace it entirely. We have covered repairing both ends of the rod, so procure a piece of soft steel rod of the right diameter and make the joints as previously explained. Do not harden the steel as it needs to be bent repeatedly throughout its life to set for beat. I have had success filling delaminated steel rods with silver solder, but it is not a guaranteed repair. If you do decide to go this route you must do the following: ensure that the delamination penetrates no deeper than halfway into the diameter of the crutch rod; thoroughly clean the split with a scratch brush and Rodico; apply flux and silver-solder as usual. If all is well, file and finish the crutch rod to disguise the repair.

Chapter 10

Reassembly

OILING

Oiling a clock movement should be done with precision and care; thought needs to be given to where and where not to place the oil, and how much to use. You will need several items to successfully oil a variety of movements: clock oil, watch oil, a couple of oil pots, and a selection of clock and watch oilers. Alternatively, you can use an automatic watch oiler for oiling platform escapements.

For clock movements you can make your own oilers. To do this, take a piece of brass or soft steel wire, the diameter of which will vary dependent on its application (1mm will suffice for an oiler suitable for most clocks), and hammer the end flat. Next, file the end into a diamond and bevel the edges. This extra step is not essential but I find it helps to direct the oil into the oil sink. If you have no brass or steel wire to hand, you can use a small screwdriver blade, dipped directly into the oil pot; the flats will pick up an amount of oil relative to the width of the blade.

For oiling platform escapements, you will need a much smaller oiler, and I suggest you buy a set of watch oilers from a horological supplier. This will provide you with a good selection for oiling pivots, pallets and escape teeth of all sizes and shapes. I personally use an automatic oiler which, at the click of a button, deposits just the right amount of oil, and for larger pivots you just click it two or three times.

You will need two oils for working on clocks, one for the movement and one for platform escapements. For the clock movement you should be using a medium-viscosity mineral oil available from horological suppliers; this will be fine for anything ranging from miniature carriage clocks up to large longcase clocks. Anything larger, such as a turret clock, will need thicker oil. Anything smaller, such as the platform escapements you will be regularly working on, will require thinner oil with different properties. When selecting a watch oil you will need to study the information provided by the suppliers and make an educated decision. Watch oils need to take account of temperatures, but in modern applications with central heating, our clocks are unlikely to experience extremes, so an oil for moderate temperature ranges should be used. Regarding viscosity, oils with a higher viscosity are intended for larger-diameter pivots such as watch barrel arbors, while the thinnest are for extreme cold temperatures where the oils will naturally thicken. I use an oil rated between these extremes, platform pivots are large when compared to watch train pivots, but are not so large as to compare to the barrel arbor of a watch.

One bottle of oil will last seemingly forever, as the amount used per oiling is so small. Do not fret about price or volume; get the best oil you can afford to save yourself from future problems – you will only buy it once a decade.

I use a small glass vial for storing the oil. Clock oil reacts to sunlight so, to protect it, paint the outside of the bottle black or wrap it in electrical tape; otherwise keep it out of sunlight when it is not in use.

Many repairers also use mainspring grease to lubricate the mainsprings. I have found that a generous application of clock oil does the trick on smaller springs, while avoiding the stickiness of the grease, which eventually dries out and can cause the coils of the spring to stick together and jump. On occasions where clock oil is not enough to adequately lubricate the mainsprings,

I use a mixture of mainspring grease and clock oil, mixed in roughly equal parts, generously applied to the spring before it is put in the barrel; any excess that squeezes out can be scraped away. I do recommend having a pot of this pre-mixed grease to hand for various applications.

Lubricating clocks is a step-by-step process which happens as you assemble the movement; try to get yourself into a routine so that you never miss a pivot. The application of lubricant will be covered during the discussion of reassembly, and I have provided you with a checklist to help build the habit.

PREPARATION

When all of the components are thoroughly cleaned, repaired and chalk-brushed, you should spend a couple of minutes getting organized before beginning to reassemble the movement. I have a divided tray on the workbench into which I place the components; the sections are arranged as follows:

- all screws, nuts and C-clips
- time train components
- strike train components
- chime train components
- all remaining 'between the plates' components.

Everything that gets fitted after the plates remains in the box until it is needed. This includes the motion works, the chime barrel, the pallets and the back cock, etc.

The first components to be assembled are the sub-assemblies: the fusee, the spring barrel, longcase barrels, etc. They require lubrication as they are assembled.

Much of the reassembly of a clock is just the reverse of the disassembly, so I have gone into detail only where notes on setting up and lubricating those components are necessary.

FUSEE

To reassemble a fusee you must first fit the gut-line to the fusee cone, unless it is a chain fusee which you fit at the very end. Cut a piece of gut-line the appropriate length for your clock, usually around about 1 metre (1 yard). Pass one end of the line through the hole in the fusee cone and tie a knot. I like to use a double stopper knot, which self-tightens as it is pulled and cannot come undone under pressure. If space permits, tuck the knot back into the provided clearance hole.

With the gut-line fitted, place the ratchet wheel on the rear of the fusee cone and screw it into place (some are a loose fit on pins to locate it centrally) being sure to get the correct orientation which is usually marked. Apply some oil to the teeth of the ratchet; a little goes a long way.

Put the click on its post on the inside of the greatwheel and apply a drop of clock oil at its centre and where the click spring contacts the tail. Finally, put a couple of drops into the centre of the greatwheel hole and fit it to the fusee arbor. To locate the click onto the ratchet, often a 'wriggle and twist' motion is enough, although sometimes you will have to press the click tail with a thin screwdriver until it clears the ratchet.

Finally, slip the key piece over the arbor and line up the screw hole, then slide it across into place. Insert the screw and double-check the fit. The greatwheel should not be sloppy, but should be able to rotate without excessive force; if not, dish the key piece slightly and try again.

SPRING BARREL

The next job is to fit the mainspring back into its barrel, and to do this a mainspring winder is required, although a hand method will be provided too. But first, for larger springs, such as fusee springs, apply a good smear of your 50/50 grease and oil mixture along the entire length of the mainspring.

Whilst wearing a thick leather glove on your left hand, fit the mainspring into the mainspring winder with the outer end pointing toward you; for chime barrels the mainspring should point away from you. Fit the clamp and wind the spring fully whilst cupping it in your left hand; this way you feel any slippage and can take action to avoid an explosive unwinding of the spring. Wind it so that the barrel will fit

Using the mainspring winder: step 1, clamp the mainspring into the winder ensuring the inner hooking eye is fully hooked and the mainspring is facing in the right direction.

Using the mainspring winder: step 2, wind the spring until it fits inside the barrel (wear leather gloves to protect your hand).

over both the spring and clamp. Make an effort to ensure that the outer hooking eye is well hooked; then, while holding the barrel tightly in your left hand, with the mainspring winders handle locked, release the clamp and pull it away from the spring. Slowly let the barrel unwind in your fingers. Finally, unhook the inner eye and remove the spring and barrel from the winder, a screwdriver is a useful aid to unhooking the spring. (If you are left-handed you may wish to use your right hand during this process.)

Back at the workbench, reach in with a pair of pliers to adjust the innermost coil of the mainspring until it is the same diameter as the barrel arbor. When the barrel arbor is fitted this coil should fit snugly against it, with the hook visible through the hooking eye. At this stage, I like to fit a winding key and, with half a turn or two, just check that the arbor is fully hooked.

Apply some oil to the mainspring at this stage. I use a bottle with a hollow needle attached to speed up the process, but repeated returns with the dip oiler will work. As you apply the oil to the side of the mainspring, capillary action will draw it down into the coils. Do not flood it, but there should be enough to thinly coat the entire surface. You will have to judge the amount yourself; if you apply too much, dab it with kitchen roll to absorb the excess.

The last thing to do is to fit the barrel cap and oil the pivot holes. Slip the barrel cap over the arbor and rotate it to the correct position, which is sometimes marked, but on fusee clocks the cut-out must line up with the three holes on the barrel. Squeeze it into place until it 'clips' in. Many are too strong to fit by hand, so I opt for a soft-jawed vice. Simply clamp the barrel between the jaws and tighten; rotate the barrel and repeat until it is closed all round.

If you do not have a soft-jawed vice you can use the handle of your hammer. Place the barrel on the edge of the workbench and, using the wooden handle of your hammer, apply your bodyweight until the barrel clips into place. Finally, apply a drop of oil to both front and rear pivots.

On a fusee clock you now need to attach the other end of the gut-line. This is done by weav-

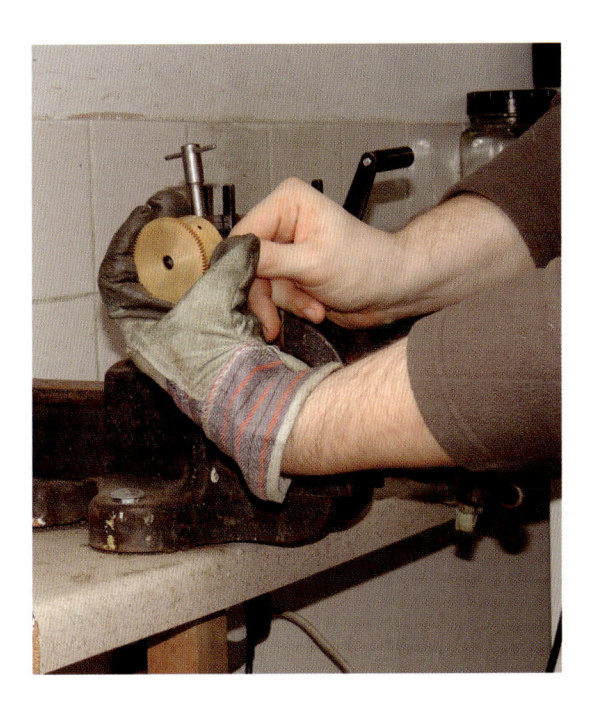

Using the mainspring winder: step 3, place the barrel over the mainspring and hook the outer eye onto the barrel hook.

Using the mainspring winder: step 4, holding the barrel tightly, remove the clamp from the outer end of the mainspring.

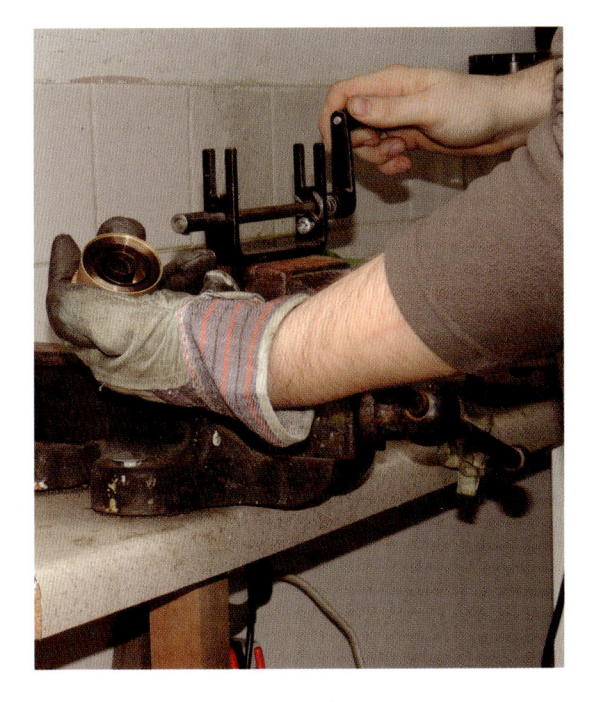

Using the mainspring winder: step 5, allow the barrel to unwind in your hand, and unhook it from the inner hooking eye.

Inner hooking eye of the mainspring fully fitted to the barrel arbor; note how the curve of the spring tightly hugs the arbor.

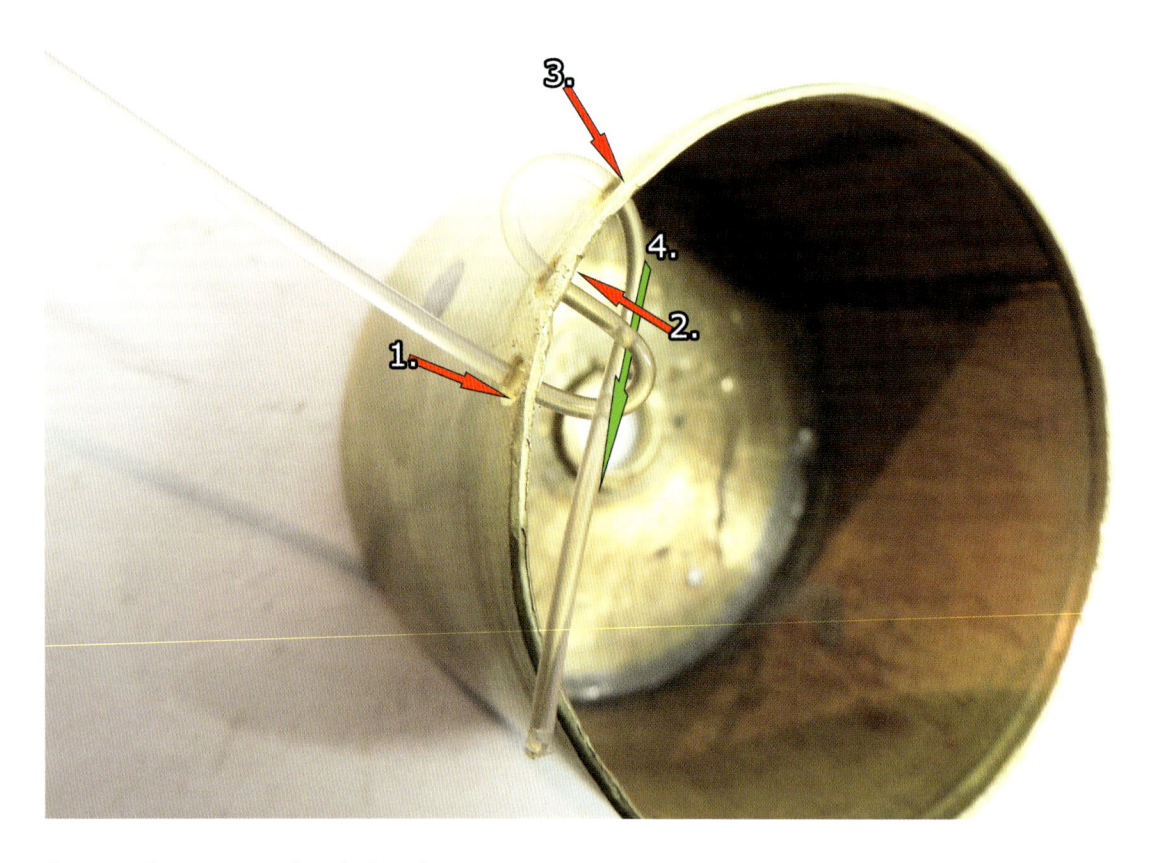

Fusee gut-line as it is tied to the barrel.

ing it through the three holes provided in the edge of the barrel as shown in the picture.

For clocks with stop work the final step is to set it up. Screw the star piece to the barrel cap with a spot of oil at its centre, and, with the barrel held in the palm of your hand, slip the finger piece over the barrel arbor but do not press it into place. Apply the winding key and give it roughly three quarters of a turn for set-up power; at this stage you need to hold the key and barrel firmly while keeping a hand free to manipulate the parts. Rotate the components so that they are about to lock into the 'fully down' position as you studied earlier, and press the finger piece into place. Slowly release the remaining power from the key and wipe any fingerprints from the barrel with a brass polishing cloth.

LONGCASE GREATWHEEL

The technique for reassembly of the greatwheel and barrel of a longcase clock is similar to that for a fusee. Pass one end of the gut-line through the small hole in the barrel and, using a wire hook, pull it through the large hole in the end of the barrel. Knot the end of the gut-line with a double stopper knot and feed it back inside the barrel.

Apply some oil to the ratchet teeth, and a drop of oil to the centre hole of the greatwheel and fit the greatwheel to the barrel. Finally, fit the key piece and pin or screw it into place. If the key piece is the pinned type, cut the end of the pin away to clear other components of the clock. Long pins here will often clash with the centre wheel or the hammer spring,

Self-tightening double stopper knot.

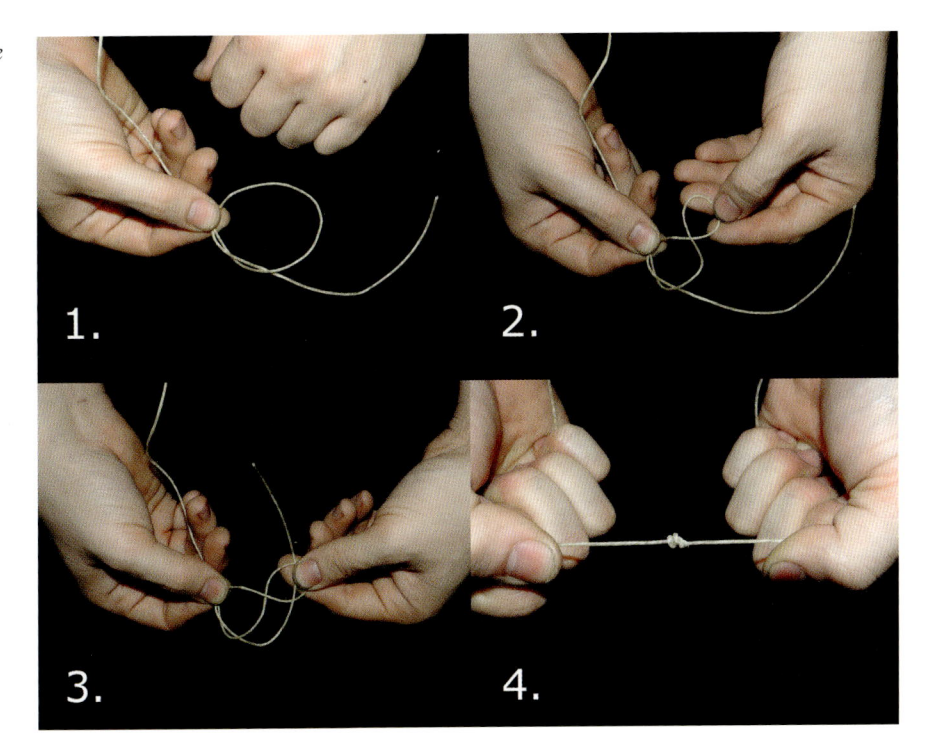

but leave enough length to the pin so that it can be pulled out by future repairers, a point which is often neglected.

THE PLATES

Next, you prepare the plates for assembly. Before you assemble the trains you should assemble the components which are screwed to the inside of the plates and which you cannot get at to assemble later. These include the hammer spring of English clocks on the rear plate and the fusee stop work and spring, which belong on the inside of the front plate. With these in place you can continue to assemble the trains.

Depending on the style of movement, attach movement holders or clock feet to the appropriate plate, as you did during disassembly, and, with the plate lying face up, begin to put the components in place. The order in which you insert the components will take some practice to work out, and taking pictures during disassembly can help speed things along. Place all of the arbors in their pivot holes, paying attention to lining up any timing marks on French movements.

For modern clocks with the friction setting at the centre wheel, put the components in place and compress the friction spring by pressing on it with pliers to clear the pin hole; with your other hand locate and insert the pin or C-clip.

For fusee movements with a gut-line, be absolutely certain that the gut-line is the right side of the pillars – it is very annoying when you assemble a clock only to realize that the gut-line is outside the pillar.

Fit the remaining plate to the movement and manipulate the pivots into their holes. By raising the working height of your workbench it is much easier to peer into the side of the movement and guide the pivots into place. Some repairers use a pivot locating tool, which can be purchased from material suppliers or made at home. I prefer to use a stout pair of tweezers which allow me to grip the pivot to locate it, as well as to lever between the shoulder and plate, raising the plate by a small amount to free a stuck pivot. It also helps to apply a small amount of downward pressure on the top plate to hold all of the arbors in place.

When you have managed to get the plates together without breaking or bending any pivots (a true testament to your newly acquired skill), lightly pin the plates together and stand the movement in the position in which it will run. You now need to set up the strike and chiming trains to their correct orientation.

To set up a rack-striking train, place the rack-hook and rack in place and rotate the train by hand. Watch as the hammer is lifted, and when it drops, hold the train steady and check for the following:

- the hammer arm should be free from all lifting pins
- the gathering pallet or locking wheel should be in its final locked position against the detent
- the warning wheel should be half a turn from the warning detent.

Often this is not the case and you need to part the plates to make adjustments. Keeping the lower pillars pinned, unpin the top pillars and reach in with the tweezers to grip the arbor to be adjusted. I will use the warning wheel as an example: hold the arbor close to whichever pivot is closest to its pinion, and lever against the plate to part them enough to release the pivot from its hole. If leverage is tight, remove all taper pins and keep the plates together with only finger tension. There should be enough freedom in the pivots to disengage the wheel and pinion. From this position, rotate the arbor to the correct position before putting it back into its pivot hole. With practice this is an efficient method of setting up the strike and chime trains of any clock.

For French clocks, you are often able to remove the pin wheel cock and completely disengage the pin wheel to line it up correctly with the timing marks on the locking wheel. If there are no timing marks then get the train into its locked position and rotate the pin wheel tooth by tooth until the hammer arm is free from its lifting pins. Unfortunately, for most English clocks this is not an option, so you must unpin the movement to rotate the pin wheel.

To set up a countwheel train, the process is the same. The hammer arm must be free of the pin wheel and there must be half a turn of warning when the locking detent has locked the train. The only remaining difficulty is adjusting the countwheel detent itself, but you will tackle this later as you assemble the rear plate.

Chime trains are often of a design which does not need adjusting between the plates as most of the components are external. If, however, you have a 'between the plates' locking model on your hands, then you will need to fit the locking levers to set up the train. Rotate the train by hand until it locks, while making a mental note of which direction the warning wheel is turning as the locking wheel rotates towards the flat surface of the locking detent. In the same manner as before, when the train is locked, part the plates enough to rotate the warning wheel until it is one quarter of a turn away from the warning detent, whose position, although you have not fitted it yet, is obvious from the cut-out in the plate.

Now all trains are set up between the plates, you can fit the taper pins, screws or nuts which hold the plates together and move on.

The Front Plate

Next, you assemble the front plate of the movement. Put the movement holders on the rear plate, with the front plate facing up, and prepare your dip oiler for use. Start by oiling each of the pivot holes for the rotating arbors of the wheels and pinions, also the hammer arbor pivot. Generally you would never oil pivots for components such as the locking detent of countwheels because, as the oil thickens with age, it will cause more problems than benefits. For now oil only the front plate. The amount of oil used should be enough so that it is visible in the oil sink, but not so much that it is full; obviously larger pivots will require more oil.

We assemble the front plate in the opposite order to which the components came off, and this varies from clock to clock. For example, for a French movement, the lifter is the first component to be fitted, as it goes beneath the cannon pinion, followed by the motion works and the remaining levers. I like to fit the click work last to ensure that there is no power on the trains while I am working on them.

Timepiece

For a timepiece, all you need to do with the front plate is to assemble the motion works. The friction setting for the motion works, if it is not between the plates, is either a friction spring (which you should dome before replacing to maintain a good friction on hand setting) or a friction cannon pinion (which rarely needs adjusting, but should be pressed on and tested for friction).

To dome a friction spring, place it on the work bench or wooden block and hammer it with a ball-pein hammer. This not only reshapes the spring to improve its tension, it also work-hardens the brass to ensure it gives a lasting result.

To tighten a friction cannon pinion there are multiple options, including an attachment for the jewelling press for very precise adjustment. For precise adjustment with minimal tooling I recommend the following technique. Pass a smoothing broach up the centre of the cannon pinion pipe. The taper of the broach will cause it to get stuck at the entrance of the pipe, while maintaining clearance towards the middle due to its taper. Squeeze the arms of the cannon pinion with a pair of pliers to close them in on the broach. When it is refitted it will be a tighter fit on the arbor, but not so tight as to be problematic.

Place the friction spring onto the centre arbor and put a drop of oil at the two highest points; if you neglect to oil this part, then with time it will cut into the cannon pinion.

Fit the cannon pinion next; this should be a loose enough fit that it slides easily over the centre arbor, followed by the bridge which supports the hour wheel if there is one, which is held in place with two screws.

For the friction fitting type of cannon pinion, press it onto the arbor until it will go no further. It is halted by a shoulder which prevents it from touching the plate and stopping the movement. Friction cannon pinions (as on French movements) have no bridge, and the cannon pinion directly supports the hour wheel.

Apply a drop of oil into the hole on the front plate, or the stud, on which the minute wheel pivots. Then smear a drop of oil onto the outside of the cannon pinion pipe, or the hour bridge pipe to lubricate the hour wheel when it is in place.

There will be no timing marks on the motion works for timepiece movements, but remember to check for them on any striking clock, especially if it is French.

Fit the minute wheel cock if there is one and screw it into place, lubricating the top pivot as you go. For a minute wheel which pivots on a post, fit the taper pin or nut which holds it in place.

Modern movements often have a washer and C-clip to hold the motion works in place; this also stops the hour wheel from falling off when you turn the movement over to assemble the rear plate. If the hour wheel is free to come off, place it to one side until it is needed so that it does not roll away unnoticed.

Strike Train

When assembling the front plate of a strike train the process is the reverse of disassembly. However, you do need to pay special attention to the fitting of the gathering pallet as this can vary from clock to clock.

Rotate the strike train by hand until it locks or so that the hammer has just dropped and the warning wheel is in its locked position, 180 degrees away from the warning detent. This position will from now on be referred to as the lock position. Depending on design, the gathering pallet will fit in one of several ways.

On English clocks, the gathering pallet locks the strike train by resting on a pin or stud on the fully gathered rack. Fit the rack and rack-hook, and with the strike train in the locked position, fit the gathering pallet so that it rests against the pin on the rack. The gathering pallet arbor is filed square as is the mounting hole in the gathering pallet, so it will only fit in one of four positions. If none of these positions lines up perfectly or the hammer is being lifted when locked, then the strike train will have to be set up between the plates again.

On French clocks, all locking takes place between the plates and the gathering pallet is pressed onto the round, tapered arbor of the locking wheel so that it is pointing directly away from the rack. Use a hollow punch or small pin chuck to press it all the way on so it cannot come loose.

For most mass-produced clocks including quarter-chimers, the gathering pallet is actually built upon a cam wheel which controls the rise and fall of the rack-hook. The rack-hook in modern cases also carries the locking detent for the strike train. Therefore the position of the gathering pallet when in the locked position must allow the rack-hook to drop into the cut-out in the cam, which will drop the locking detent into the path of the warning wheel, locking the train.

Do not oil the pivots of these levers, but a thin smear of grease is recommended on all surfaces which are in frictional contact. For example, the lifter is in frictional contact with the lifting pin on the minute wheel or cannon pinion, so apply a drop of grease on the pin. While you have the grease pot out, put a drop on a few lifting pins on the pin wheel, which will spread evenly during use. Use clock oil on the warning detent because grease is too thick and will stick to the warning pin.

With the front plate assembled you should check that the motion works is set up correctly, or else the minute hand will not set off the strike on the hour. Fit the minute hand and turn it to see at which point the lifter drops from the lifting pin. The hand should be pointing perfectly top and centre to the movement. If the pins are part of the cannon pinion this problem cannot occur unless the cannon post is twisted or the pins bent, but if the lifting pin is on the minute wheel and the strike is not being released in the right position, do the following:

- remove the pin, nut or cock which holds the minute wheel in place
- turn the minute hand until the lifter drops
- disengage the minute wheel to cannon pinion gearing while holding the minute wheel in place
- turn the minute hand to point at twelve o'clock
- re-engage the minute wheel and cannon pinion
- re-test and repeat as necessary.

Chime Train

The final part of the front plate to set up is the chime train, and in most cases this is easy to do. Start by fitting all of the levers in place in the reverse order to which they came off, followed by the locking cam which is slipped onto its arbor. Tighten the grub screws enough to provide friction but allow it to slip around when forced.

Rotate the train in the direction it works until the warning pin is a quarter-turn away from the warning detent, which is the locked position. The locking cam will have found its locked position and be resting hard against the locking detent. If not, complete one more full rotation of the warning wheel, then tighten the grub screws fully.

In many chiming trains you can hold their position by inserting a pair of tweezers in the crossing of the warning wheel through the cut-out which is intended for fly adjustment. This makes setting up the train much easier.

Fit the quarter countwheel to its arbor, being sure to get the correct orientation by noting in which direction the arbor rotates, and in which direction the countwheel must rotate. You may need to lift the self-correction levers to fit it all the way on the arbor. Finally, adjust the depth of the fitting to allow clearance for all of the surrounding components. For example, the self-correction lever for the chime should rest on the cam at the rear of the quarter countwheel, but should not be able to clash with the countwheel when all end-shakes are taken into account. Tighten the grub screws enough to provide some friction.

Rotate the countwheel so that in the locked position the locking detent rests in the gap between the first (smallest notch) and second (second smallest notch) quarter on the countwheel. You will see why you choose this position later. Tighten the grub screws fully.

Finally, apply a drop of oil to the warning detent, and thinned grease to all components in frictional contact, such as the lifter and the pin which raises the locking detent. Pin or fit C-clips to all components to that they cannot fall off as you turn the movement over.

The Rear Plate

Fit the movement holders to the front plate of the movement now and turn it face down on your workbench. Just as you did with the front plate, lubricate all of the pivot holes with clock oil, including the hammer arbor, but not the countwheel-detent arbor or any other lifting arbors which extend through the plates.

If the clock has a rear countwheel this should be the first part you fit as it is usually beneath the hammer and crutch. Note the timing mark on one face of the square on which it fits, and the corresponding timing mark on the countwheel. Fit it in the indicated position and pin it in place. If there is no timing mark, then you will have to try all four positions to see which works best.

Test the strike by providing power by hand. It is not uncommon that the countwheel locking detent has become worn or been 'adjusted' in the past, which can cause problems. You have already corrected the wear groove during repair and all that is left is to readjust the detent until the locking is correct.

Release the strike repeatedly, counting the number of blows at each step of the countwheel. It should be sequential, with a single blow between each full strike for the half-hour. The detent should lock in the gaps of the countwheel and not on the slope. Try the countwheel in all four positions until this is correct, unless the train is marked.

When detents lock on the slope of the countwheel it is because it is not lifting high enough at this point to clear the locking detent, common if a wear groove has been filed away or it has been previously adjusted. To correct this, slightly twist the detent downward. There can be so many possible adjustments here that a bit of patience and thinking time is needed to make the right choices.

With the countwheel in place it is time to fit the anchor. (Come back to this step later if you are repairing a fusee clock.) Feed it through the provided hole and into position, fit the back cock and screw it into place. Put some power on the train by hand and watch the action of the escapement. Most back cocks can be slightly knocked up or down to adjust the drops, but check first that the back cock is not pinned. If you do need to adjust the drops, leave the screws tight and use a block of wood resting on the back cock, tapped lightly with a hammer to knock it up and down as necessary. Lubricate both the front and rear pivots with clock oil, and place a drop of oil on each pallet face. Power the going train by hand to work the oil around the escape teeth and repeat the process.

Hammers are simply slipped onto their arbors and pinned into place.

CHIME BARRELS

Reassembling the chime train is just the reverse of disassembling. However, they do need to be correctly set up.

If you remember, you set up the quarter countwheel so that it is locked having just chimed the first quarter. This is because Westminster and Whittington chimes complete a descending tune at quarter past which is visible on the chime barrel as a straight run of pins, at a slight angle along the barrel.

With the chime barrel and all of its auxiliary components reassembled onto the plates, lubricate with clock oil the studs on which any wheels are pivoted; leave loose the adjustable wheel on the arbor extension through the rear plate.

Rotate the chime barrel by hand so that it lifts and releases the hammers while watching for the straight run of hammers that symbolizes the first quarter. At the end of this run, tighten the grub screws on the adjustable wheel, just enough so that they grip the arbor, but can be forced to slip by hand. The chime barrel should now be in sync with the quarter countwheel.

Put power on the chime train by hand and lead the minute hand round until the chimes release. Watch the action, and if any hammers are left in the air as the chime train locks, rotate the adjustable wheel until it drops. Perform the same test on each tune by turning the selector knob or lever. When all hammers are clear of the chime barrel and the tune is showing a straight run at quarter past, fully tighten the grub screws on the adjustable wheel.

Unfortunately, not every barrel is assembled to show a straight run of hammers on the quarter hour, and this is obvious by studying the gongs, which should be arranged by length. If the gongs are not arranged by length then you will not be able to set up the chime barrel in this manner, and will have to leave the adjustable wheel a friction fit until the movement is back in the case and you can set it up by ear rather than by eye. The method is the same, but listen for a straight run of descending notes rather than looking for a straight run of hammer blows.

Finally, apply grease to the pins of the chime barrel, and oil the chime barrel pivots. Do not oil the hammer pivots as this will cause problems in the future as the oil thickens.

PLATFORM ESCAPEMENTS

With the platform escapement cleaned and the pivot holes pegged, their reassembly is fairly simple, but it is important that they are properly lubricated along the way so I will detail the process step by step.

With the main plate of the platform upside down on a boxwood ring or watch movement holder, start by oiling the lower jewel-holes of the balance. I use an automatic oiler and three clicks of the button dispenses the right amount of oil for most platforms, but if you are using a dip oiler, select an oiler large enough to fill the pivot hole without flooding the jewel. Make an effort to keep all of the oil in the centre of the jewel, or else it will be drawn away over time and the lubrication will fail.

Place the end stone into position and screw it down. The oil should form a circle in the centre of the end stone. Some people prefer to place a drop of oil in the centre of the end stone before putting it in place instead of in the pivot hole. The oil will then be drawn into the pivot hole in the right quantity.

Oil the lower pivot of the escape arbor in the same manner. Never oil the lever pivot; the viscosity of the oil slows down the action of the lever, stealing power from the balance wheel. Turn the main plate over on the movement ring.

Fit the escape wheel and guide it into the lower pivot hole, also fit the lever unless you are working on a cylinder escapement.

Fit the bridge or separate cocks for the escape wheel and lever. Use a piece of pegwood to apply very gentle downward pressure on the cock as you guide the pivot into the hole with the tweezers. This can be a fiddle, but do not be tempted to use any force to complete this step.

Fit the screws, but before you drive them all the way, check the end-shakes of the arbors to ensure that they are located in the pivot holes. Drive the screws all the way.

Lubricate the top pivot of the escape wheel but not the lever pivot. Place the main plate to one side to finish later.

Place the upper end stone of the balance assembly upside down on the work surface, and then fit the index over it. Oil the pivot of the balance cock and lower it into place on top of the end stone, being careful to line up the screw holes. Place a piece of pegwood through one screw hole to locate it relative to the hole beneath it, and rotate the entire cock until the other hole lines up; fit one screw, and then the other.

Finally, open the 'boot' of the index and guide the balance stud into its hole, while guiding the hairspring between the index pins. With a screwdriver or flat-faced punch, push the stud into its hole, and tighten the screw if there is one. Close the boot of the index around the hairspring.

Lift the balance assembly by the index and turn it over. With the main plate back on the movement holder, guide the balance into its lower pivot hole, either around the cylinder or so that the impulse pin is between the horns of the lever. Fit the balance cock and guide the top pivot into its hole.

Before you tighten the balance cock screw, lightly blow on the balance wheel to set it oscillating. If it stops as you tighten the screw then it is not properly located and you should immediately release the screw and start again.

Finally, oil three or four tooth tips of the escape wheel and, while putting light finger

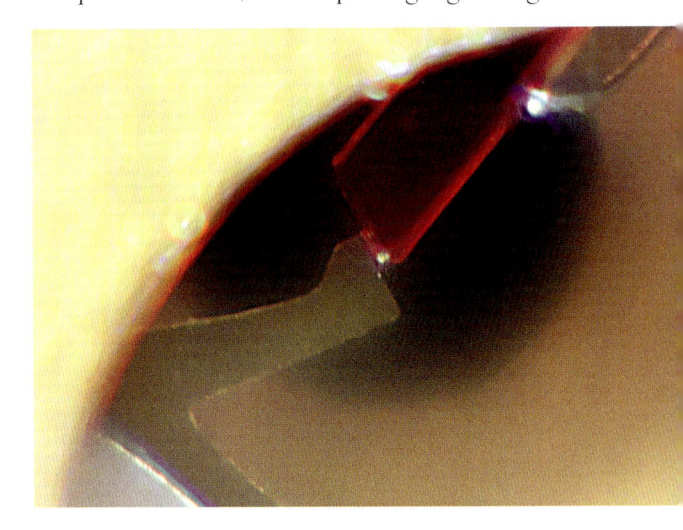

Correct amount of oil on the pallets of the lever escapement.

pressure on the escape pinion, set the escapement running to spread the oil. The result should be that a minute amount of oil is visible between the tooth tips and the lever pallets when viewed under strong magnification.

Fitting the platform to the movement should be the final assembly job after refitting the click work. Fit all four screws loosely and move the platform back and forth until the depthing of the contrate wheel appears right: unfortunately, finding this position by eye takes experience, and only testing will show its true position. Tighten the platform screws to hold it in place.

CLICK WORK

The final piece of the puzzle before testing is to assemble the click work. This does not apply to fusee clocks.

The easiest way to do this is to fit the click spring and click in place first, placing a drop of oil on the shoulder of the click screw and where the click spring and click come into contact.

Slip the ratchet wheel over the winding square and into position, bearing in mind that they are not always interchangeable. If the ratchet wheel is too tight or loose for the arbor, then try swapping them over.

The ratchet wheel will be held in place by either a cross pin through the winding square, or a cock and screw. Once this is in place, lubricate a few teeth of the ratchet wheel with clock oil or thinned grease.

FUSEE SET-UP

When the front and back plate are assembled it is time to set up the fusee and fit the ratchet.

Screw the click loosely to the plate, and pin the ratchet wheel in place. Fit a large winding key or T-square to the barrel arbor square (not the fusee winding square) and rotate the barrel until the hook hole is exposed or the gut-line is completely unwound.

Lightly oil the fusee chain to prevent it from rusting in the future and hook the spear-shaped hook to the barrel hole while keeping some tension on it, or the gut-line, as you wind the barrel arbor. This will not wind the spring as the whole barrel is free to turn.

Wind the gut-line or chain onto the barrel, guiding it to follow the marks made by years of gut-line or chain. When you reach the end of the chain, place the click into the teeth of the ratchet and keep tension on the chain. Rotate the fusee with the winding square and key until the hook pin is visible, hook on the rounded hook of the chain and take up the slack by winding the barrel square. The train will turn as the chain pulls at the fusee until it is in its fully unwound position.

With the chain or gut-line fully wound around the barrel and hooked at both ends you can set up the fusee spring.

Turn the barrel arbor by half a turn. This time the spring will be wound because the line is holding the barrel in place. Fit the click into the ratchet and release your grip slowly; the click should hold back the power of the spring. Re-grip the winding key and wind the barrel to provide final set-up power.

Set-up power was explained in the theory section of this book and is individual to each spring, as is the fusee cone. As many clocks are no longer on their original springs, or the properties of that spring have changed over time, it is not necessary to calculate the amount of set-up power necessary. Simply estimate the point at which the power of the spring seems to increase during winding and stop there. This is usually around the half- to three-quarter-turn point.

Guide the click into the teeth of the ratchet as you release the winder, and finally tighten the click screw.

Wind the movement, guiding the chain or line into the groove of the fusee as you go. When the clock is fully wound (if you cannot fully wind it then you have too much set-up power or need to replace the mainspring), let it run down completely before continuing. This will allow the line to find its natural position on the barrel, as well as helping to spread the grease and oil around the mainspring.

When the clock is fully down, double-check the set-up power now that the line has settled. Finally, you can fit and lubricate the anchor as described earlier.

Chapter 11

Tips on Testing

Accurate testing of the clock movement is one of the most important defining points between a clock which 'runs' and a clock which actually works as it should.

Firstly, you will need a number of test stands, which you can make at home or buy at horological auctions. Plans for acceptable test stands can be found in the resources section at the rear of this book.

Initial testing should be done without the dial but with the hands fitted. Testing without the hands allows for too much end-shake in the motion works, which may cause the strike to jam or the wheels to disengage, as well as retracting the ability to test the timekeeping of the clock.

Good notes should be kept for all clocks on test, which is easy to overlook, so I have designed the job card (provided in the resources section) to help prompt us to keep good notes.

The provided test card has seven sections all containing space for the same information. Initially when we set a clock on test, we note down the date on which it was wound and set, denoted by 'W/S'. When the clock stops, we report the reason why – for example, if the rack-tail jammed on the snail, or the hands were touching each other. We then wind and set the movement once again and note the date. This process can be repeated until the movement completes a full cycle (eight days usually) of perfect running; we are also able to spot recurring patterns easily to diagnose a fault.

The clock should be perfectly set up for testing; that is, in beat, upright, set to time and stable. You are trying to mimic how the clock will be set when it is in its final home. The test stands are designed to allow this and to not interfere with the movement.

A sturdy shelf should be provided for testing; a clear wall should be available into which you can insert screws or, as I have done, screw a strong peg board, onto which you can hang clocks in nearly any arrangement, and finally you will need a sturdy floor on which to place longcase test stands. Wobbly floorboards mean that the clock can be disturbed every time you walk past. If you are uncertain, screw the test stand to the wall.

When you are convinced that the clock has run a full cycle with no problem, the following will confirm that you are ready to move onto the next stage:

- the countwheel and hour hand are in sync
- the clock has kept time to within five minutes per week
- on winding, all trains are down by a similar number of turns (although bear in mind that repeating clocks have a gear ratio to allow for additional striking of the hour so should require less turns of the key)
- longcase and Vienna regulator weights have dropped at the same rate.

If all of the above points are true, then you can proceed to the next step, fitting the dial and testing once again.

The dial is fitted in the reverse order to which it was removed, and if it is part of the case, as is common with 1930s mantel clocks, then just fit the movement back into the case at this stage.

I like to test the clock with the dial on for a few days before fitting it to the case, repeating the testing process from before and making notes along the way. If all is well, fit the movement to the case to complete the week of testing. In situations where the case was left with the customer (as with longcase and Vienna regulator clocks), test the movement with the dial on for the remainder of the week.

After two weeks of perfect running, the clock is ready to return to its owner.

Chapter 12

Setting Up

THE IMPORTANCE OF SETTING UP

The setting up of a clock is the final part of the repair process and it needs to be done properly. The process is slightly different for each type of clock but follows a few general rules:

- clocks, especially those with a pendulum, need a stable base upon which to be mounted
- they need to be in beat or they will eventually stop; this could be after two seconds or two hours depending on just how out of beat they are
- they need to be wound
- they need to be upright, and the pendulum needs to be free to swing uninterrupted.

COMMON BEAT-SETTING METHODS

Give the pendulum a push start and listen to the tick while watching the swing. If the tick is off to either side you need to adjust the crutch.

To set up a friction fit crutch, common in modern clocks and antique French clocks, carry out the following actions. If the tick is off centre to the left, then push the crutch to the right; eventually the anchor will bury itself into the root of the escape wheel and stop rotating, which is felt as increased resistance as the anchor begins to slip on its arbor. With a bit of trial and error, you can get the beat position on centre. It takes practice but soon you will be able to get results quickly. New clocks are often sold with the intention of the customer making this adjustment by themselves.

To set an antique English clock in beat, support the crutch at the top and bottom with the fingers of one hand, and bend the middle of the rod toward that hand with the fingers of the other. The resulting bow in the crutch will shift the beat position, so recheck it after every attempt. This is done with the movement fully assembled. Always support the crutch in this manner to avoid breaking it.

If the tick is off to the left, then the left hand should play the supporting role; if it is off to the right, then the right hand forms the support.

When the clock is in beat, check that the crutch foot is not twisted to one side and that the suspension block is in or near the centre of the crutch foot and not at one extreme.

LONGCASE CLOCKS

When setting up a longcase clock you should approach the job systematically. Get it stable, get it upright, wind it and get it in beat.

Stability

The case should ideally be screwed to the wall. Many of my customers have had their clocks standing for years well away from the wall and on carpet which is on top of wobbly floorboards. This may have worked in the past, but you can nearly guarantee that if you leave it like this, there will be problems and your time will be wasted. In this situation I never guarantee that a clock will run until the stability situation is corrected.

When the weights of a longcase reach the same level as the pendulum, they will swing in harmony and eventually the entire case will rock. This rocking robs the pendulum of impulse and eventually the clock will stop. It is important to eliminate the rocking, and the best way to do this is with a screw through the case and into the wall.

Most longcase clocks have holes in the rear of the case where they were once screwed in place, so there is usually no need to make new ones. If the wall has a skirting board, it is important that you make up for the additional thickness by placing a block of wood between the case and wall and passing the screw through it.

If the customer does not wish to fix a screw to their walls it is necessary to find a way round the problem. In this situation, a selection of wooden wedges and blocks is essential. Push the base of the case hard against the wall and find a block of wood the right thickness to place behind the case near the top so that it is level when pushed backwards against the wall. Use wooden wedges beneath the front feet to lean the clock hard against the wooden block. The friction of the clock pressing against the wall will be enough to keep most cases stable. If in doubt, add a little Blu-tack to the wooden block.

Verticality

Sit the seat board of the movement on the cheeks of the case and hang the pendulum. Longcase movements are very front-heavy and have a tendency to fall forwards if not weighted down. The dial should be upright from all angles; I like to set them using a plumb line from twelve to six so they are as upright as possible. Use wooden wedges between the cheeks and the seat board to achieve this.

Winding the Clock

The gut-lines should be completely unwound in the workshop before you return the clock. While pulling down on the line to keep it tight, wind one train of the clock halfway and hang the weight. Repeat the process for the other weight. When both weights are hung, wind the clock fully.

Spring-driven longcase clocks just need to be wound as usual.

Refit the hood and check how the movement lines up in the opening. Remove the hood and adjust the position of the movement left or right to centre the dial in the hood. This should be done before setting the clock in beat, and the weights being in place help support the movement as the hood is removed.

Swing the pendulum and set the clock in beat using the appropriate method explained earlier.

Finally, set the time, allowing the clock to strike each hour as you go, also set any date work or moon dials before you fit the hood. It is a good idea to hang around for five minutes to be certain all is well.

DIAL CLOCKS

Dial clocks are set in beat prior to returning them to the customer, and this is best done on the test stand. Set the movement upright on the appropriate test stand using a plumb line to check the verticality of the twelve and six o'clock markers.

Physically setting the clock in beat is the same process as for most antique English clocks, by bending the crutch rod, and finally ensuring that the pendulum is free on the crutch foot.

My final piece of advice is to make sure that the clock is hung upright, by placing a spacer between the clock and the wall. Also be certain that a strong screw sunk deep into the wall is being used, and not a brass picture hook.

MANTEL CLOCKS

Mantel clocks require a stable and level base to sit on, whether it is a mantelpiece or a shelf. This means that you should avoid placing them on a wobbly unit, or on a stable unit on a wobbly floor; any vibration passed to the clock as people walk past could be enough to disturb it. Any case with a short foot, which is not uncommon, should have a piece of felt or cork placed beneath it to stabilize it. A wobbly case will rock with the pendulum and cause stoppage.

Life is made a lot easier if the clock is far enough away from the wall that one can open the rear door and reach in to adjust the pendulum, but it is not essential.

Most of the common 1900s mantel clocks have a friction crutch.

FUSEE BRACKET CLOCKS

Fusee bracket clocks are similar in their requirements to most mantel clocks, but the beat-setting method is typical of English clocks, by bending the crutch.

VIENNA REGULATORS

Vienna regulators should be hung from a strong wall screw because they are weight-driven and can be heavy. Hang the case so that it is upright from all angles and fit the pendulum and movement.

If the pendulum rubs the back board, make sure that the suspension spring is not stuck in its slot at an angle. If it is then adjust this before proceeding to tilt the case forward to free the pendulum.

With the gut-lines fully unwound, check that the end loop of the gut is properly hooked in place, and keep tension on the lines while winding the clock. When it is wound just enough that it will clear the bottom of the case, hang the weight and repeat the process for the other line. Leave the weights toward the bottom of the case to create the clearance needed to set the clock in beat.

Reach up to the bottom of the crutch, where the foot is mounted to a thread and can be adjusted left and right. This has the same effect as bending the crutch rod. Wind the knurled nut to move the crutch foot left or right until the beat of the clock is even. As always, check that the pendulum is not resting on the crutch rod.

Wind the clock and set the hands, allowing it to strike each hour. Wind until the weights are just below the dial. Go too high and the pulley will ride up and over the knot of the loop and the weight may fall off and damage the case.

Chapter 13

Aftercare

Mechanical clocks require regular maintenance to keep them in good order and to time. This is the clock owner's job and not the repairer's, so you will need to provide instructions on how it is best done.

Regular maintenance includes adjusting, regulation, winding, dusting, etc., as well as arranging regular maintenance with the repairer every three to five years. Below is a sample letter, which you could provide for the clock owner so that they are aware of their responsibilities toward the clock, now that yours have ended. The customer is fictitious and should be replaced with your own.

FOR THE ATTENTION OF: A CLOCK OWNER

Dear Mrs A. Fictitious,

Now that your clock has been fully overhauled by us, you will be pleased to know that we offer a full one-year guarantee from the date of its return. Unfortunately this offer cannot extend to breakages and mainsprings, as there are factors involved here which are out of our control.

Your clock is now expected to run like new for a period of about ten years; however, this will only happen with your active involvement in maintaining it.

Your responsibilities toward the clock now that ours have ended are: regulation, setting, winding, adjustment of the strike and cleaning the case regularly. You should also schedule your clock for professional attention every three to five years.

All clocks leave our workshop keeping time to within five minutes a week, but you should be able to achieve better timekeeping when final regulation is done with the clock in its final position.

Regulation
To regulate your pendulum clock you will need to adjust either the rating nut on the pendulum or the adjuster on the dial. If yours is a French clock, you will need a second key which we can supply for a small cost. To slow a clock which is gaining, you need to lengthen the pendulum. This is done by unscrewing the rating nut in increments of about half a turn at a time or by turning the dial rating square to the left. Lower is slower. The opposite is true for a clock which is losing time.

For balance-controlled clocks including carriage clocks, the index lever is located at the top of the movement and is nearly always labelled 'S' and 'F' for slower and faster, or 'R' and 'A' for retard and advance. Simply push the index toward the intended result in increments of about 1mm at a time.

Do not be tempted to adjust your clock more often than every three days to account for power fluctuations and temperature errors.

Winding
Never be afraid to wind your clock fully. This is easier with weight-driven clocks because you can watch the weight rise in the case and stop winding as it reaches the top. For spring-driven clocks you will need to wind each square until there is a sharp increase in resistance, signifying that the clock is fully wound. Do not try to wind it any further; this is when spring damage can occur. Fusee clocks should be wound until a physical stop is reached.

You should wind all squares or chains, otherwise problems could occur with the strike mechanism jamming the clock.

Strike problems

If your clock has countwheel striking, it is likely that it will at some point get out of sync with the hand position. Countwheel striking has always been liable to this and was superseded by rack striking because of it. On the rear of the clock, the countwheel should be visible; it is a brass disk with cut-outs or pins around its edge. Lift the small lever which is sitting in one of the cut-outs on the circumference. This will set the strike running. Repeat this process until the number of blows matches the position of the hour hand. If it is a recurring problem, return the clock to be checked over by us.

Case cleaning

If you have a brass-cased clock, my best advice is to keep the dust off and use a brass polishing cloth. Most brass and gilded clocks are lacquered and should not be polished with brass polish or any other abrasive.

Wooden clocks should be regularly dusted and periodically waxed with a natural beeswax, and the brass finials should be wiped down with a brass polishing cloth. Any other style of case should be dusted regularly and otherwise left alone.

Routine maintenance

All that is left is to schedule regular maintenance every three to five years. This maintenance is charged at our hourly rate and in most cases is no more than an hour's work.

Large clocks have greater oil reserves and can be left up to five years without attention in most cases, while small carriage clocks will dry out much more quickly and should be oiled after three. After six years, it is advisable to have the platform escapement overhauled; this is two hours of work but will reduce potential troublesome wear further down the line.

If you keep to this schedule, it is entirely likely that your clock will last more than ten years between overhauls.

Final tips
- Never move a clock with the pendulum in place.
- Never spray-oil the clock in any circumstance.
- Stop the pendulum if you go on holiday.

We would like to thank you again for choosing us to service your clock(s). If any issues arise, we would like to be the first to know.

Many thanks.

Chapter 14

Resources

RECORD CARD

The double-sided card shown on this and the facing page is what I use in my own workshop, but you may wish to add other items or vary the size to fit your own needs:

Name:
Phone:

Date recieved:
Date completed:
Date collected:
Collected: X

Name:
Phone:

Date recieved:
Date completed:
Date collected:
Collected: X

Date recieved:

Instructions:

Description:
Condition:

Name:
Address:

Home:
Work:
Mobile:
Email:

Estimate:

Date recieved:

Instructions:

Description:
Condition:

Name:
Address:

Home:
Work:
Mobile:
Email:

Estimate:

Name:
Date: ETA:
Estimate (+/-15%):
Type:
Condition:

Full overhauls of any clock are guaranteed for one year from collection. 'Part jobs' eg. suspension, mainspring only etc. can be carried out as requested, but **will not** be guaranteed without full overhaul.
By accepting this ticket you accept our terms.

In the event of any query this reciept must be produced.

Name:
Date: ETA:
Estimate (+/-15%):
Type:
Condition:

Full overhauls of any clock are guaranteed for one year from collection. 'Part jobs' eg. suspension, mainspring only etc. can be carried out as requested, but **will not** be guaranteed without full overhaul.
By accepting this ticket you accept our terms.

In the event of any query this reciept must be produced.

Front page of my record card system; the bottom part stays with the clock owner, the middle is our record and the top stays tied to the clock.

Back page of my record card system; the repair card in the middle is laid out for simple repair record keeping.

212 Resources

WORKROOM EQUIPMENT

The diagrams provided are not precise plans to be followed to the letter. The measurements given are of the equipment I use, or improvements of that equipment. Use stock material to save time and use these diagrams as a guide for tried and tested designs. I firmly believe in the use of modern adhesives in making practical test rigs, and simple joints make for much quicker and more repairable work.

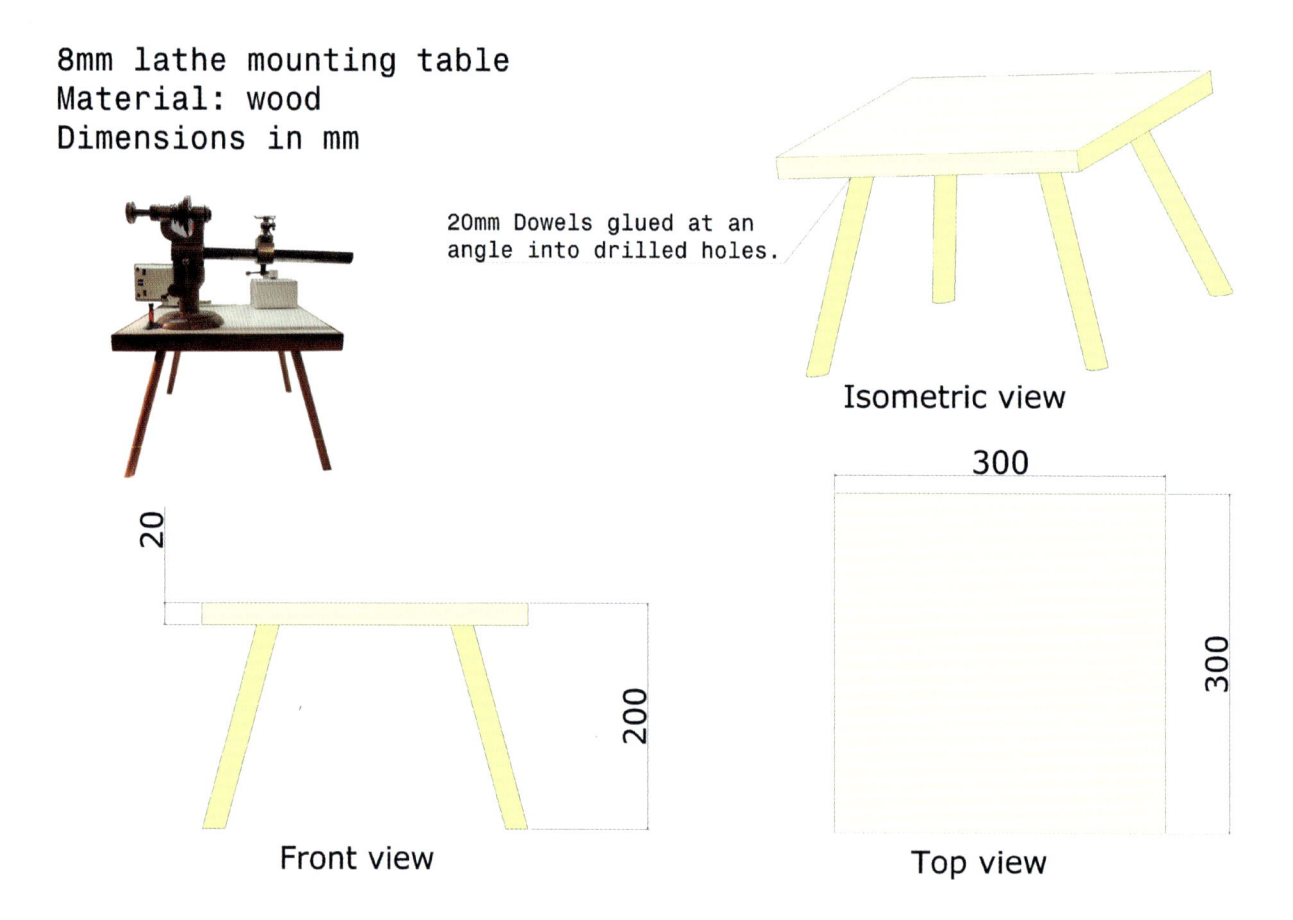

8mm lathe mounting table
Material: wood
Dimensions in mm

20mm Dowels glued at an angle into drilled holes.

Isometric view

Front view

Top view

Watchmaker's lathe mounting table; this is the table I use (the dimensions are not critical).

Raised work surface
Material: Plywood
Dimensions in mm

10
10
10mm Lip around table top.
Dowels glued into holes.
200
600
372

Raised height workbench; this sees regular use in my workshop when assembling clock plates.

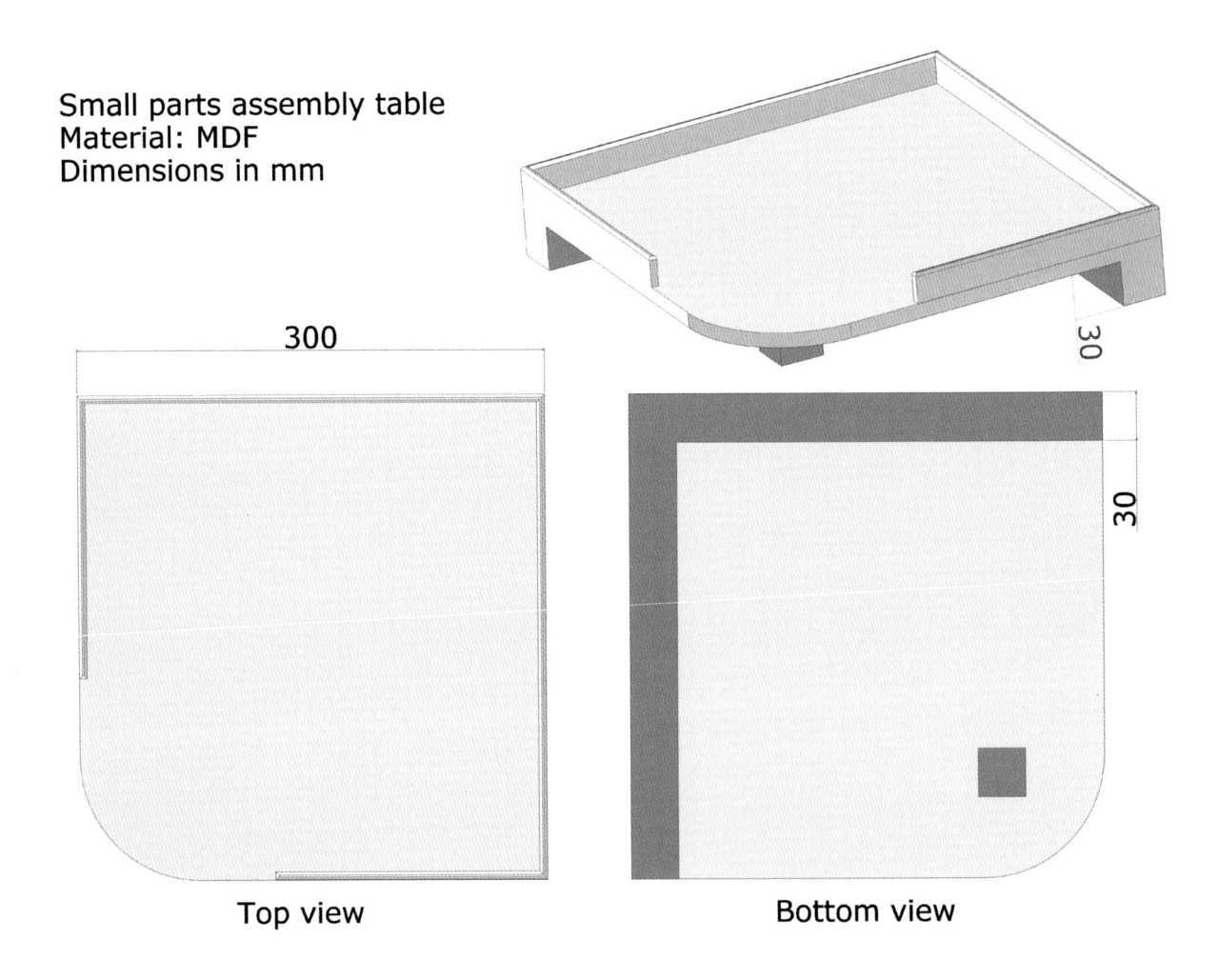

Small parts assembly table
Material: MDF
Dimensions in mm

300

30

30

Top view

Bottom view

Platform repair table I use when repairing small parts; the design is good for catching any components that try to escape.

French movement test stand.
Material: Hardwood
Dimensions in mm

Countersink the nut.

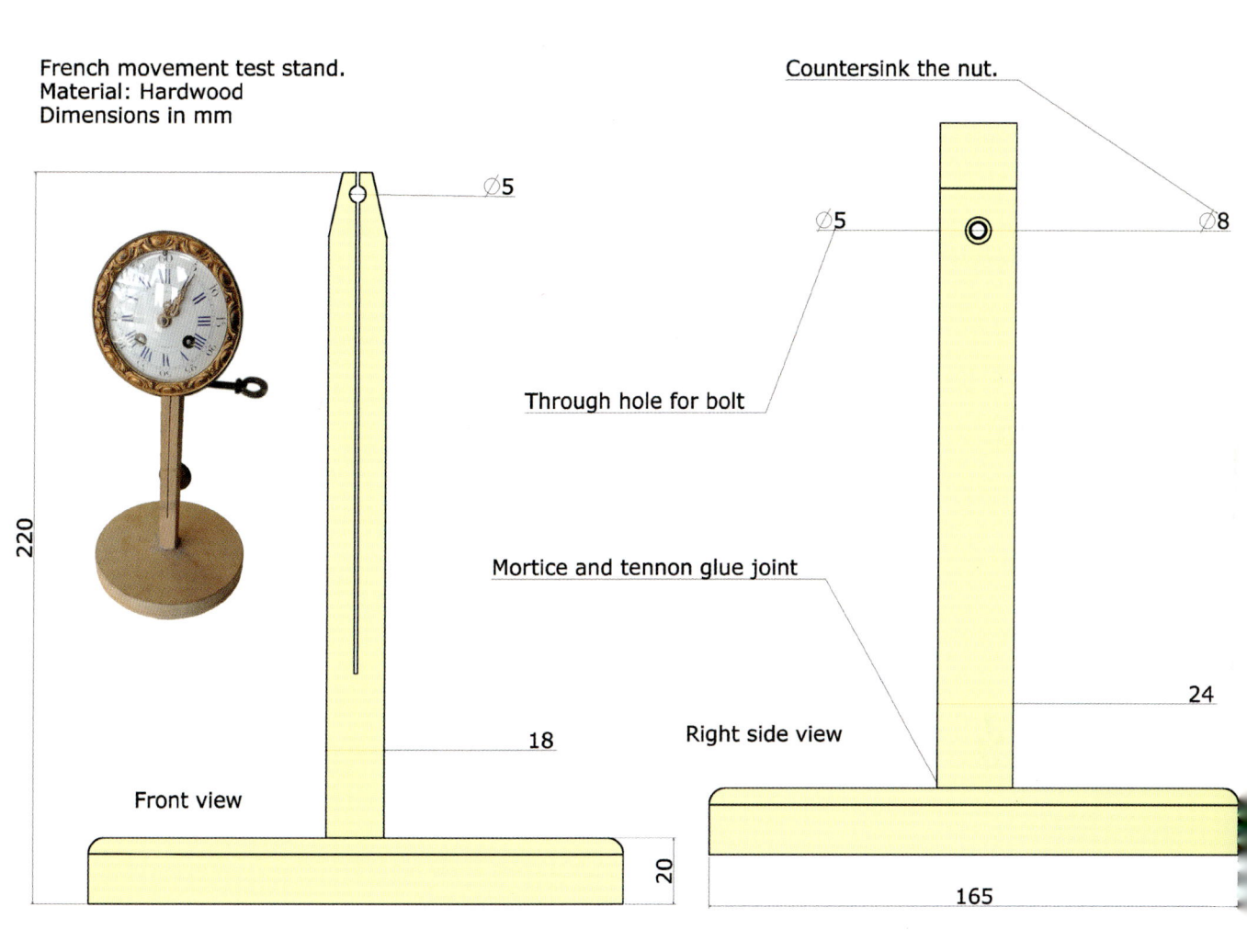

Ø5

Ø5

Ø8

Through hole for bolt

Mortice and tennon glue joint

220

18

Front view

Right side view

24

20

165

Test stand I use for testing French clock movements: a bolt is needed to pass through the provided hole, and the nut should be a tight fit in the countersink to stop it from rotating.

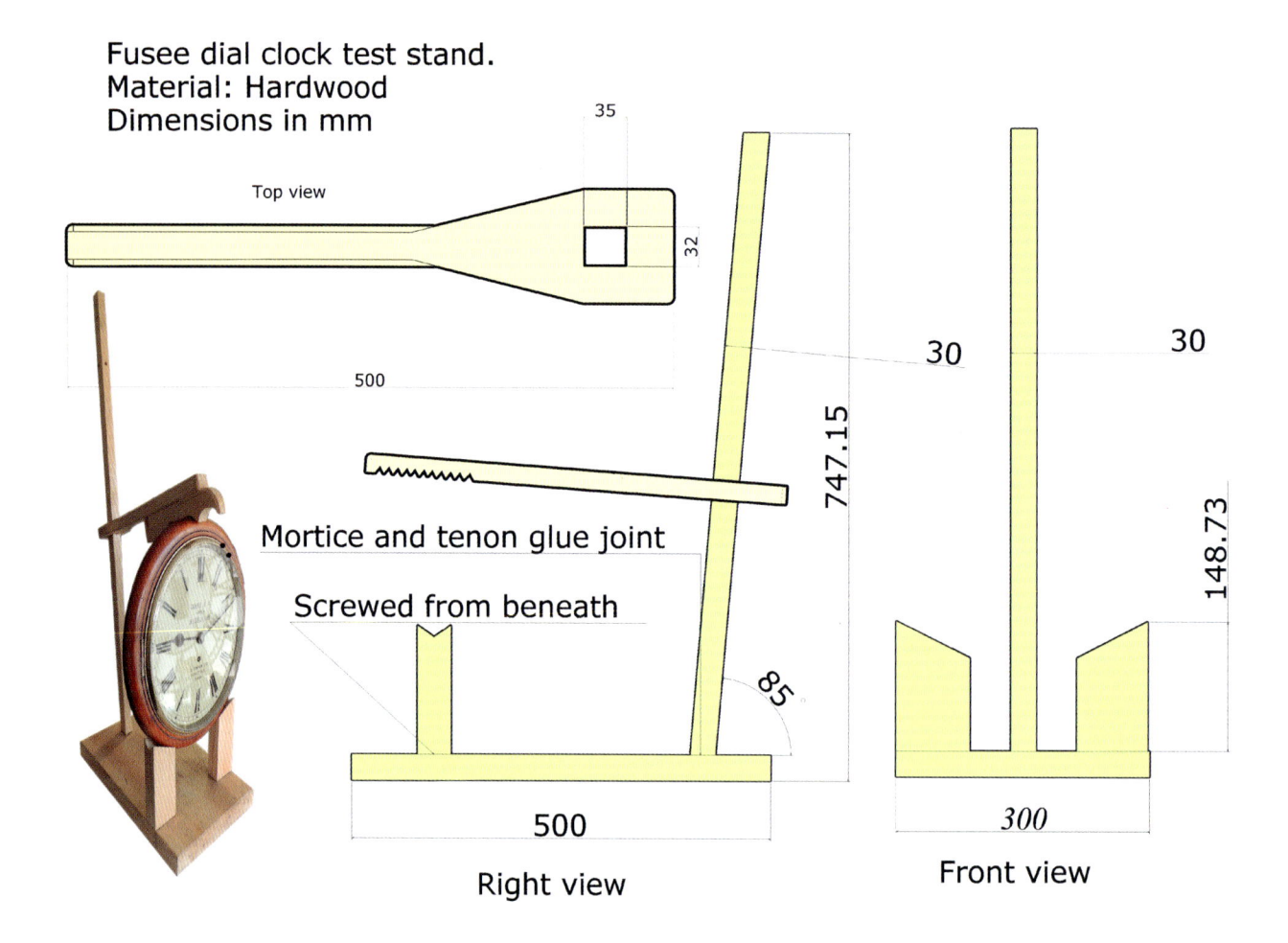

Fusee dial clock test stand.
Material: Hardwood
Dimensions in mm

Top view

35
32
500

Mortice and tenon glue joint

Screwed from beneath

30
30
747.15
148.73
85
500
300

Right view

Front view

Fusee dial clock test stand; the clock bezel is mounted by clamping between the test stand and the arm (this design can accept most dial and drop dial clocks).

Resources 217

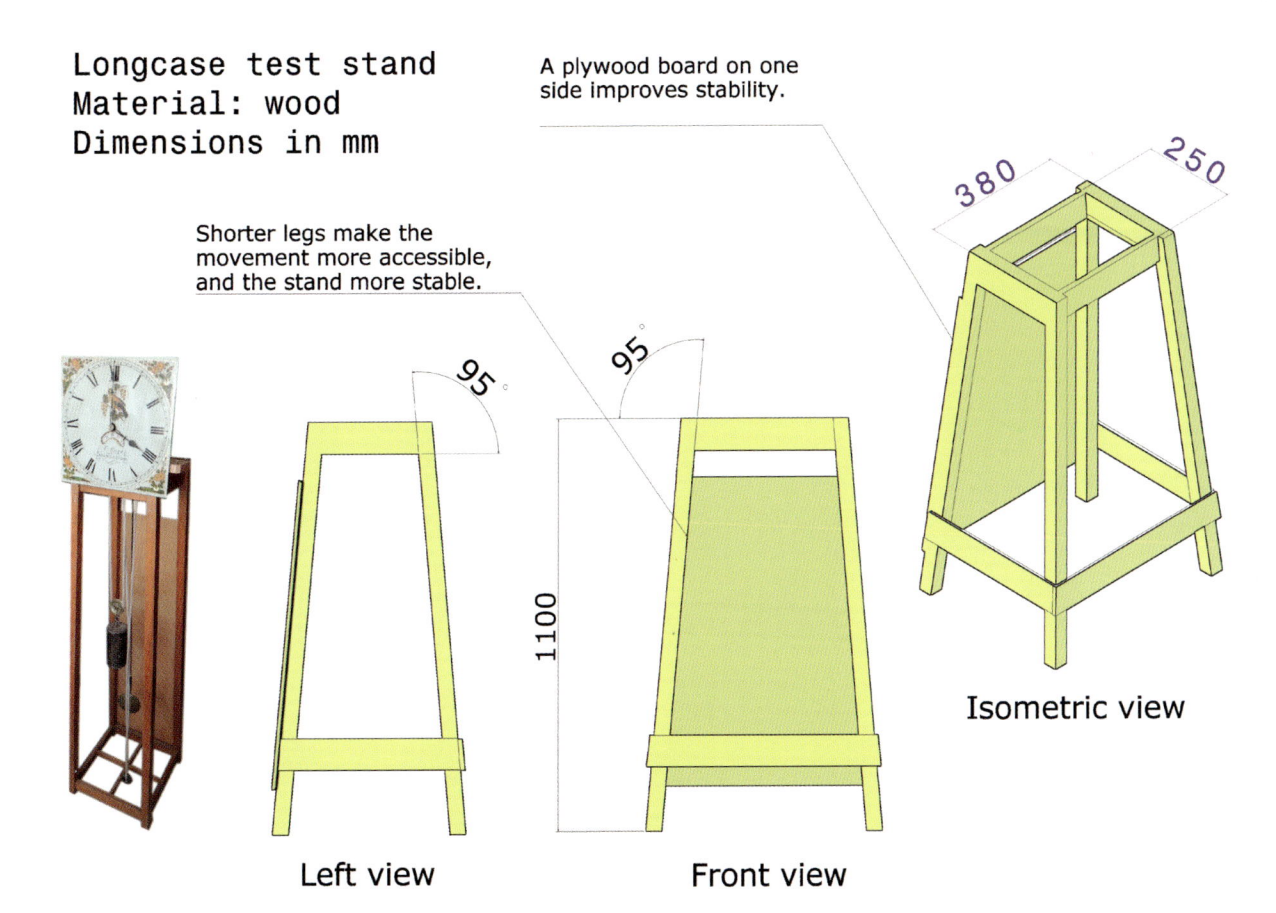

Longcase test stand
Material: wood
Dimensions in mm

A plywood board on one side improves stability.

Shorter legs make the movement more accessible, and the stand more stable.

95°

95°

380

250

1100

Left view

Front view

Isometric view

Longcase test stand, an improvement on the design I use; the legs have been shortened for improved stability and accessibility; and the top has been lengthened to accept larger movements.

218 Resources

Vienna regulator test board
Material: plywood
Dimensions in mm

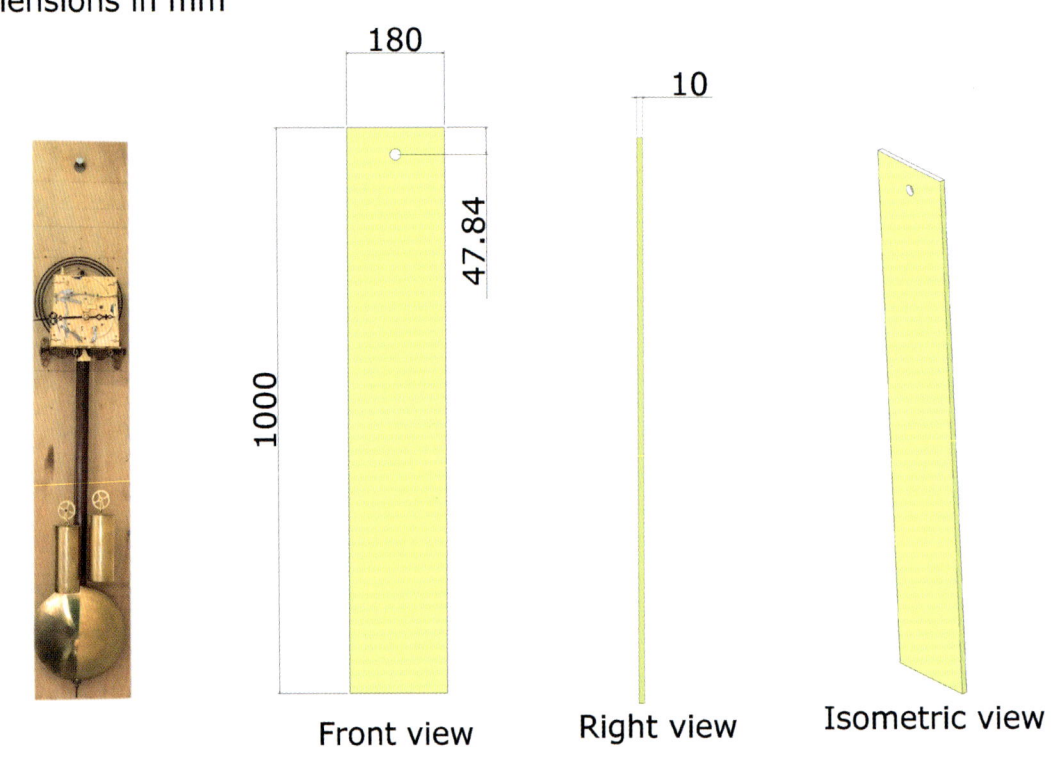

180

10

47.84

1000

Front view   Right view   Isometric view

Vienna regulator test board; cut from a piece of plywood, this test board will accept any Vienna regulator and hangs from a wall screw.

CLOCK REPAIR CHECKLIST

1. Initial inspection:

- Are the hands touching?
- Is the clock wound?
- Is the clock in beat?
- Is the case stable?
- Is the strike jammed?

2. Transportation:

- Have you checked for a hood latch?
- Have you removed the key from the top of the clock?
- Remove the hood.
- Remove the weights.
- Remove or lock the pendulum.
- If possible, lift out the movement.
- Unscrew the suspension block if possible.
- Wrap all parts in bubble wrap to avoid scratches.

3. In the workshop:

- Remove the hands.
- Remove the dial.
- Let the power out of the mainsprings.
- Remove the click work if possible.
- Remove the platform escapement.
- Check the pallet arbor pivots.
- Remove the pallet arbor and check for wear.
- Strip the front plate, checking for wear.
- Check all pivot holes for end- and side-shake.
- Strip the plates.
- Check the pivots for scoring.
- Strip the platform escapement.
- Remove and inspect the mainsprings.
- Perform all repairs.

4. Cleaning:

- Remove thick deposits with pegwood.
- Remove all rust with emery paper or the scratch brush.
- Clean the components ultrasonically or by hand.
- Rinse twice.
- Dry thoroughly.
- Chalk-brush all parts.
- Pegwood all pivot holes.
- Polish out all severely etched fingermarks.
- Reassemble the fusee or spring barrel, remembering to lubricate as you go.
- Reassemble the plates.
- Check for end-shake and missed pivot wear.
- Set up the strike or chime trains for locking.
- Lubricate the front plate as you reassemble.
- Pin all front plate components in place.
- Lubricate the back plate.
- Assemble the back plate, including the escapement.
- Fit the hands for testing.

5. Testing:

- Fit the movement to the appropriate test stand.
- Wind it up.
- Hang the pendulum and set it in beat.
- Set the hands to time.
- Synchronize the countwheel to the hands.

6. Setting up:

- Stabilize the case.
- Assemble the clock.
- Wind it up.
- Set it in beat.
- Set the hands, checking the strike as you go.
- Synchronize the countwheel.
- Double-check:
    – the hands are not touching
    – the clock is in beat
    – the strike is correct
    – the case is stable.
- Advise the clock owner of their aftercare responsibilities.

Glossary

Arbor A shaft or axel.

Amplitude The amount of rotation in degrees, of the balance wheel.

Anchor A component of the recoil escapement which carries the pallets.

Anniversary clock The 400-day anniversary clock, often given as gifts for birthdays and anniversaries as they only require winding once per year.

BA British Association screw threads, often used in clockwork.

Balance The oscillator of a watch or platform escapement.

Banking pins Used to limit the motion of the lever escapement.

Beat The tick of a clock, an 'in beat' clock will have an even and regular tick.

Boring tool A cutting tool used for cutting the inside surfaces of stock in the lathe.

Bridge A supporting plate containing one or more pivot holes, fixed at each end.

BTM Lathe 8mm watchmaker's lathe of British manufacture.

Burnishing Hardening and polishing the surface of a material.

Butterfly A type of knot common in clock making.

Cannon pinion The first gear in the motion works.

Cock A supporting plate containing one or more pivot holes, fixed at only one end.

Collet A work holding tool for the lathe.

Complication Any mechanical addition to a clock beyond timekeeping; e.g. date-work is a complication.

Dead centre (object) A tool that marks the exact centre height of the lathe.

Dead beat An escapement without recoil.

Detent A physical stop that prevents motion of the train until released .

Draw A design feature of the lever escapement that holds the lever against the banking pins.

Drop A necessary but wasteful component of any escapement to allow freedom of the escape wheel.

End shake The end-to-end freedom of an arbor between the plates.

Entry pallet The first pallet on which an escape tooth will land as it enters the escapement cycle.

Exit pallet The last pallet, on which an escape tooth will exit the escapement cycle.

Escapement The regulating mechanism of a clock.

French four-glass clock A style of clock with a glass panel to all four sides of the case; the movement is often of the standard French design.

Fusee A conical brass component used to smooth the power output of a mainspring.

Graver Hand-cutting tool for the lathe.

Greatwheel The first gear in a train.

Guard pin A safety feature of the lever escapement.

Hairspring A fine spring used to regulate the oscillations of a balance wheel.

Headstock The working end of a lathe.

Heat sink A large metal object used for removing heat.

Horology The science of clocks and watches.

Impulse The part of the escapement cycle that provides power to the pendulum.

Jewel Synthetic jewels used to reduce friction and wear in clocks and watches, often in the form of pivot holes.

Knocking cut When the surface that is being turned on the lathe is not continuous and 'knocks' against the tool as it makes and breaks contact, for example when turning the corners from a square rod; this can be detrimental to the life of the tool.

Mainspring A common power source for small clocks.

Movement The complete working mechanism of any clock or watch.

Oscillator The pendulum or balance wheel.

Pinion A small steel gear that is driven by the larger wheel.

Poising Removing heavy spots from a wheel, often by filing to improve power transmission.

Pivot The thinned end of an arbor, on which the gears rotate.

Reamer Cutting tool used to accurately open a hole to sizes.

Seat board The board to which the movement is mounted.

Shellac An excretion of the lac beetle, used as an adhesive.

Side shake The side to side freedom of a pivot in its hole.

Staff A balance wheel's arbor.

Stop work A power-smoothing device used to limit the winding and unwinding of the mainspring.

Train A series of gears.

Index

aftercare 208–9
anvils 16
auxiliary repairs 175–82
    bells 179–80
    hands 175–8
    pulleys 180–82

balance wheels 62
ball-pein hammer 10
barrel hooks 136–8
barrel teeth 153–5
barrels 105–7
bells 74–5
bench knife 18
bench vice 17
bent pivots 123–4
bent teeth 144–6
blowtorch 19
brass 109
brass bushes 16
bristle brush 17–18
broaches 15
brocot pallets 190
brocot pin pallet escapement 56–7
broken teeth 149–52
burnisher 14–15
bushing 116–
bushing wire 16

callipers 23
carbon springs 39
chalk 17–18
    brushing 113–14
chime barrels 85
    reassembling 201–2
chime train 83–4
chiming clocks 82–5
circular error 60
cleaning 109–15

click and screw 141–3
click spring 144
click work 42–3, 203
clock oil and oiler 11
clock repair checklist 219
clock theory 30
clockmaker's butterfly 36–7
contrite gearing 48
countwheel 75–7
countwheel train 78–9
crutch 50
crutch rod 190–91
cutting tools 16
cylinder escapement 63–6

deadbeat anchor escapement 53–5
depthing tool 20–21
dial clocks 206
diamond files 18
dies 26
disassembly 87–107

end shake 120
endless loop 35
escapements 49–73

files 24–5
flux 10
front plate 84
fusee 14, 39, 40, 50, 96, 107, 193, 201
fusee bracket clocks 207
fusee set-up 203

gear depthing 45–8
gear theory 44–9
Geneva stop work 41–2
getting started 88–107
glossary 220–21
gongs 74–5

Index

hacksaw 25–6
hairsprings 62–3, 185–6
hand cleaning 113–14
hand vices 15
hooking eye repair 131–3

jewelling press 21
jewelling tool 21

lantern pinions 165–6
let-down tool 13
lever platform escapement 66–73
line of centres 47
longcase clocks 205–6
longcase greatwheel 196–7
loose fly 168–9
loose pillars 173
loupes 13

machine vice 17
mainspring repairs 134–6
mainspring winder 14
mantel clocks 206
materials 108–9
micrometer 23
motion works 48–9
movement holders 11

needle files 11

oil 11
oil sink cutting tools 16
oil sinks 16
oiler 11
oiling 192–3

pallet wear 187–9
parting the plates 103–5
pegwood 18
pendulum theory 58–61
piercing saw 25
pillar drill 24
pin chucks 15
pinion wear 160–64
pivot file 14–15
pivot shape 67
plates
    reassembling 197–201
    stripping 100–103
platform escapements 61–73, 96–9
    reassembling 202–3
platform repairs 183–7
    bushing platforms 185
    cylinder wear 186–7
    hairsprings 185–6
    jewel holes 183–4
pliers 9
polishing 114–15
power source 33

rack 75–7
rack-striking train 79–81
rack-tails 167–8
rags 13
ratchet wheel 141
re-finishing pivots 120–22
re-pivoting 125–30
rear plate 85
reassembly 192–203
recoil anchor escapement 50–53
record card 210–11
record keeping 87
repairs 116–91
resources 210–19
riveting 175
rubbed jewel tool set 21–2

scratch brush 18
screwdrivers 8–9
setting up 205–7
    longcase clocks 205–6
    common beat-setting methods 205
    dial clocks 206
    fusee bracket clocks 207
    mantel clocks 206
    Vienna regulators 207
silver solder 27
smiths little torch 27
solder 10
soldering 174–5
soldering iron 10
split pin holes 170–72
spring barrel 193–6
spring-driven clocks 35–
stainless springs 39
stakes 16
staking set 22–3
steel 108–9
stones 24–5
stress corrosion cracking 31

striking clocks 74–81
   Dutch 75
   Roman 75
   ship's 75
   trumpeters 75
stripping the plates 100–103

taper pins 33
taps 26
teeth
   barrel 153–5
   broken 149–52
   wear 147–8
test stands 11
testing 204
timepiece, the 30–73
timing marks 81
tin snips 19
toolboxes 8–27
   toolbox 1 8–13, 116, 139, 141, 144, 173, 185, 186, 190
   toolbox 2 13–19, 116, 120, 131, 144, 160, 165, 168, 180, 187
   toolbox 3 19–27, 123, 125, 134, 136, 141, 145, 147, 149, 153, 156, 167, 170, 175, 179, 183, 185, 190
tooth wear 147–8
transportation 88–91
truing callipers 21
turning 173–4
tweezers 13

ultrasonic cleaning 111–13
ultrasonic cleaning tank 23

vices 17
Vienna regulators 207

warning 77
watch-cleaning machine 23
watchmaker's lathe 19–20
wear grooves 116
weight-driven clocks 33–5
wheel, fitting a new 156–9
winding square 139–41
workbench 29
workroom equipment 212–18
workshop design 28–9